Springer Theses

Recognizing Outstanding Ph.D. Research

Aims and Scope

The series "Springer Theses" brings together a selection of the very best Ph.D. theses from around the world and across the physical sciences. Nominated and endorsed by two recognized specialists, each published volume has been selected for its scientific excellence and the high impact of its contents for the pertinent field of research. For greater accessibility to non-specialists, the published versions include an extended introduction, as well as a foreword by the student's supervisor explaining the special relevance of the work for the field. As a whole, the series will provide a valuable resource both for newcomers to the research fields described, and for other scientists seeking detailed background information on special questions. Finally, it provides an accredited documentation of the valuable contributions made by today's younger generation of scientists.

Theses may be nominated for publication in this series by heads of department at internationally leading universities or institutes and should fulfill all of the following criteria

- They must be written in good English.
- The topic should fall within the confines of Chemistry, Physics, Earth Sciences, Engineering and related interdisciplinary fields such as Materials, Nanoscience, Chemical Engineering, Complex Systems and Biophysics.
- The work reported in the thesis must represent a significant scientific advance.
- If the thesis includes previously published material, permission to reproduce this must be gained from the respective copyright holder (a maximum 30% of the thesis should be a verbatim reproduction from the author's previous publications).
- They must have been examined and passed during the 12 months prior to nomination.
- Each thesis should include a foreword by the supervisor outlining the significance of its content.
- The theses should have a clearly defined structure including an introduction accessible to new PhD students and scientists not expert in the relevant field.

Indexed by zbMATH.

Zheng-Yang Huo

Nanowire-assisted Flow-through Electrode Enabling Electroporation Disinfection of Reclaimed Water

Doctoral Thesis accepted by
Tsinghua University, Beijing, China

清華大学出版社
TSINGHUA UNIVERSITY PRESS

Springer

Author
Dr. Zheng-Yang Huo
School of Environment and Natural Resources
Renmin University of China
Beijing, China

School of Environment
Tsinghua University
Beijing, China

Supervisor
Prof. Hong-Ying Hu
Environmental Simulation and Pollution Control State Key Joint Laboratory,
State Environmental Protection Key Laboratory of Microorganism Application and Risk Control (SMARC),
School of Environment
Tsinghua University
Beijing, China

ISSN 2190-5053 ISSN 2190-5061 (electronic)
Springer Theses
ISBN 978-981-99-4501-6 ISBN 978-981-99-4502-3 (eBook)
https://doi.org/10.1007/978-981-99-4502-3

Jointly published with Tsinghua University Press
The print edition is not for sale in China. Customers from China please order the print book from: Tsinghua University Press.

This Springer imprint is published by the registered company Springer Nature Singapore Pte Ltd.
The registered company address is: 152 Beach Road, #21-01/04 Gateway East, Singapore 189721, Singapore

Supervisor's Foreword

Disinfection can efficiently inactivate microorganisms and control biological risks in water quality, but existing disinfection technologies have associated risks such as difficulty in controlling by-products and easy microbial reactivation. In this thesis, the nanowire electroporation disinfection technology was developed based on the flow-through treatment mode of porous electrodes. The topic selection has important theoretical significance and application value.

The main innovative achievements of this paper are as follows:

(1) The internal filtration nanowire electroporation disinfection technology has been developed, which has improved the efficiency of microbial disinfection, achieved low-voltage (1.0 V) safe disinfection, and significantly reduced the potential of disinfection by-product generation in the treatment process.
(2) The nanowire electroporation disinfection technology can realize the inactivation of bacteria without reactivation, and can promote the gradual death of damaged bacteria during storage.
(3) The use of high-frequency alternating current power supply mode can ensure the disinfection effect of electroporation, inhibit the occurrence of electrochemical reaction, and effectively prolong the service life of electrodes.

The thesis has a reasonable structure, advanced methods, rich content, rigorous discourse, detailed data, standardized writing, and reliable conclusions. This outstanding work indicates that the author has a solid and broad theoretical foundation and in-depth professional knowledge in the field of this discipline, and has the ability to independently conduct scientific research.

Beijing, China
May 2023

Prof. Hong-Ying Hu

Foreword

Water is a basic human need. The increasingly severe water scarcity has made wastewater reclamation a necessity. During wastewater reclamation, disinfection is an essential process to protect people from pathogen infections. However, conventional water disinfection methods (e.g., chlorination and ultraviolet radiation) suffer from obvious drawbacks such as harmful by-products formation and pathogen regrowth. Thus, there is an urgent need for an efficient and safe wastewater treatment technology that can achieve safe and efficient pathogen disinfection. Electroporation, a physical process relying on a strong electric field to damage microorganism membranes, can be a promising substitute to reduce the chemical demand of conventional water disinfection methods. In this study, a flow-through nanowire-assisted porous electrode was fabricated to achieve efficient pathogen disinfection in wastewater reclamation with low voltage and within several seconds of contact time. This study provides theoretical guidance and technical support for controlling biological risks in reclaimed water.

The copper-oxide-nanowire-modified foamed copper electrode prepared by the thermal oxidation method enables the safe and efficient disinfection in the reclaimed water at an external voltage of 1 V and hydraulic retention time (HRT) of 10 s. The extremely low external voltage ensures that no Cl_2 is produced during operation, which reduces the potential for the formation of harmful disinfection by-products (DBPs). Attributed to the typical disinfection mechanism, no bacteria growth occurs during the storage process afterward.

The aspect ratio of the nanowire is the core factor affecting the disinfection effect. When the aspect ratio is greater than 40, the complete microorganism disinfection can be achieved. In addition, reducing the positive electrode aperture to 200 μm while maintaining the negative electrode aperture (500 μm) can reduce the electrode cost while ensuring the disinfection efficiency.

Replacing copper oxide (CuO) nanowires with cuprous phosphide nanowires can effectively improve the stability of nanowire-modified electrodes. The electrode lifetime increases 36 times from 20 min to 12 h, making it suitable for most small water treatment facilities. Attributed to the enhanced stability, the electrode degradation rate decreases by one order of magnitude, and the effluent copper concentration decreases from 350 μg L^{-1} to 25 μg L^{-1}. Furthermore, the electrochemical reaction of the electrode surface during the long-term operation is the main cause of electrode degradation.

The polydopamine (PDA) film is uniformly deposited on the electrode surface, which can effectively improve the electrode mechanical stability and reduce the electrochemical reaction at the water-electrode interface. Coated with PDA film, the stability of the electrode is further enhanced. The electrode lifetime increases to 4 days, and the electrode degradation is significantly reduced with a lowered copper concentration in the effluent (14 μg L^{-1}). Furthermore, the application of a high-frequency altering current (AC) power output (10^6 Hz) enables electroporation and avoids electrochemical reactions simultaneously. Under such operation conditions, the PDA-coated Cu_3P-nanowire-modified copper foam electrode achieves complete disinfection for 15 days continuously, and the copper concentration in the effluent is only 4 μg L^{-1}.

Nanowire-modified electroporation disinfection technology can achieve complete disinfection for secondary effluent from municipal wastewater treatment plants and reclaimed water receiving waters with an external voltage of 1 V and HRT of 10 s. The quality of reclaimed water had no significant effect on the electroporation disinfection performance.

Beijing, China
December 2020

Dr. Cheng Tang

Parts of this thesis have been published in the following journal articles:

1. Huo, Zheng-Yang, et al. Synergistic nanowire-enhanced electroporation and electrochlorination for highly efficient water disinfection. Environmental Science & Technology, 2022, 56(15), 10925-10934.
2. Huo, Zheng-Yang, et al. Evaluation and prospects of nanomaterial-enabled innovative processes and devices for water disinfection: A state-of-the-art review. Water Research, 2020, 173, 115581.
3. Huo, Zheng-Yang, et al. Elevating the stability of nanowire electrodes by thin polydopamine coating for low- voltage electroporation-disinfection of pathogens in water. Chemical Engineering Journal, 2019, 369, 1005-1013.
4. Huo, Zheng-Yang, et al. Low-voltage alternating current powered polydopamine-protected copper phosphide nanowire for electroporation-disinfection in water. Journal of Materials Chemistry A, 2019, 7(13), 7347-7354.
5. Huo, Zheng-Yang, et al. A Cu3P nanowire enabling high-efficiency, reliable, and energy-efficient low-voltage electroporation-inactivation of pathogens in water. Journal of Materials Chemistry A, 2018, 6(39), 18813-18820.
6. Huo, Zheng-Yang, et al. Cell transport prompts the performance of low-voltage electroporation for cell inactivation. Scientific reports, 2018, 8(1), 1-10.
7. Huo, Zheng-Yang, et al. Impact of water quality parameters on bacteria inactivation by low-voltage electroporation: mechanism and control. Environmental Science: Water Research & Technology, 2018, 4(6), 872-881.
8. Huo, Zheng-Yang, et al. Carbon-nanotube sponges enabling highly efficient and reliable cell inactivation by low- voltage electroporation. Environmental Science: Nano, 2017, 4(10), 2010-2017.
9. Huo, Zheng-Yang, et al. Nanowire-modified three-dimensional electrode enabling low-voltage electroporation for water disinfection. Environmental science & technology, 2016, 50(14), 7641-7649.

Acknowledgments

I sincerely thank Prof. Hong-Ying Hu, my Ph.D. supervisor, for his sincere teaching over the years. He cultivated my good research taste and taught me what kind of research is the best. Under his guidance, I have experienced the far-sightedness and meticulous logic of academic masters, which has benefited me throughout my life. Peach and Plum are speechless, and I have forged my own unique path. Professor Hu's words and deeds over the years have been precious to me throughout my life.

Thanks to Prof. Xie Xing and Prof. Liu Nian from the Georgia Institute of Technology in the United States for helping me with my project. Thanks to Dr. Ding Yong for her careful guidance on material characterization. Thanks to Prof. Lu Yun, Prof. Wen Xianghua, Prof. Lu Wenjing, Prof. Jiang Kaili, Prof. Wang Jiaping, Prof. Wei Jinquan, Prof. Wu Hui, Prof. Wu Qianyuan, and Prof. Yang Cheng from Tsinghua University for their unreserved discussions and selfless help on the subject.

Thanks to all my colleagues in the research group. Every day I work with them is precious.

Thanks to my parents and family for their constant encouragement, support, and understanding.

Thanks to my wife Gao Yue.

Contents

Chapter 1
Introduction

1.1 Research Background

1.1.1 Significance of Wastewater Reclamation and Reuse

Water is a basic human need, whereas one in three people do not have reliable access to safe drinking water worldwide every year [1]. The contaminated drinking water has become one of the main transmission pathways of waterborne pathogens leading to diarrheal, malaria, and gastroenteritis [2, 3]. Thus, waterborne illnesses remain a huge threat to public health, especially in the developing areas where sanitation facilities and gird power are not available or in developed areas that suffer from natural disasters (e.g., earthquakes and tsunami) [4, 5]. More specifically, as a representative of developing countries with the most population, China is a severely water-deficient country. The per capita water resource in China is only 2300 m^3, which is 25% of the world's average level [6]. At the same time, there are serious water pollution problems in China, which have caused poor quality water and further exacerbated water shortages [7]. In 2016, more than 23.1% of rivers were polluted, including 9.8% of severely polluted rivers. The pollution of lakes is even more severe, with 17.8% of lakes severely polluted [8]. In light of the current severe water shortage problem, it is of great significance to develop unconventional water sources to support traditional water sources to relieve the shortage of water resources.

Wastewater reclamation has become an important way to solve the current challenge of water shortage in developing countries [9]. Given that urban sewage has the characteristics of easy access, stable water quality, and low cost of reuse, wastewater reclamation and reuse can be low cost and high efficiency, which has become an effective unconventional way to supplement water resources [10, 11]. In addition, in the process of wastewater reclamation, the discharge of pollutants can be reduced, effectively [12]. At present, wastewater reclamation and reuse have been widely applied in urban use, agricultural irrigation, industrial water, and drinking

Z. Huo, *Nanowire-assisted Flow-through Electrode Enabling Electroporation Disinfection of Reclaimed Water*, Springer Theses,
https://doi.org/10.1007/978-981-99-4502-3_1

reuse in China [13, 14]. At the same time, other countries have successfully applied the deep reuse of wastewater to direct drinking reuse. China's 13th Five Year Plan clearly stated that wastewater reclamation and reuse should be applied as an effective way to supplement urban water sources. The Plan pointed out that by the end of 2020, Beijing's wastewater reclamation rate should reach 68% and the wastewater reclamation rate of other provinces and cities should strive to reach 15% [13, 15]. Wastewater reclamation has broad practical engineering applications, whereas the current large-scale and in-depth wastewater reclamation raises the concern for higher safety requirements.

1.1.2 Necessity of Wastewater Reclamation and Reuse

Pathogenic microorganisms exist inevitably in municipal wastewater and may cause various diseases and pandemics [16, 17]. Pathogens in the wastewater have the characteristics of multiple biological species, high concentration levels, and high infection risks. Thus, the control of pathogens has become a key issue that needs to be paid attention to during wastewater reclamation [18, 19]. Table 1.1 summarizes the concentration and risk of various pathogenic microorganisms in the undisinfected effluent from municipal wastewater treatment plants.

Based on the infection probability of one single exposure of pathogens, the World Health Organization (WHO) calculates the infection risk of pathogens in reclaimed water. Several pathogens showing a high infection risk exit in reclaimed water. Disinfection can inactive pathogenic microorganisms and control the biological risks of pathogens in the process of wastewater reclamation, effectively. It has become an indispensable process to ensure the safety of the reclaimed wastewater [31].

1.1.3 Challenges of the Existing Disinfection Technology

During wastewater reclamation, microbial disinfection has become the essential method to control the biological risks and ensure water safety. Table 1.2 summarizes the conventional disinfection technologies applied in wastewater reclamation. Chlorine disinfection has become the most widely used water disinfection technology attributed to its low operating cost, good microbial disinfection effect, and constant disinfection performance. However, the harmful carcinogenic disinfection by-products (DBPs) produced in the chlorine disinfection process will involve new health and ecological risks, inevitably [32–34]. In addition, the disinfection effect of chlorine disinfection on protozoa (i.e., giardia and cryptosporidium) is relatively limited [35]. Emerging disinfection technologies such as UV disinfection can effectively control the generation of DBPs during the operation process and effectively inactivate the pathogenic microorganisms in water. However, after UV disinfection, the inactivated bacteria will regrow or reactivate [36, 37], and UV disinfection has

Table 1.1 Concentration and risk of pathogens in effluent of municipal wastewater treatment plants before disinfection

Pathogens	Concentration	Potential risk	References
Bacteria	(count/100 mL)		
Enterococcus faecalis	$900–10^4$	Intra-abdominal infection, urinary tract infection	[20]
Clostridium difficile	$50–10^5$	Diarrhea	[21]
Salmonella	$\sim 1.0 \times 10^2$	Food poisoning	[22]
Clostridium perfringens	$\sim 9 \times 10^3$	Enteritis, food poisoning	[23]
Pseudomonas aeruginosa	$\sim 1.1 \times 10^4$	Lung infection	[24]
Viruses	(copy/mL)		
Adenovirus	$10^3–10^5$	Respiratory tract infection, gastrointestinal infection, urinary tract infection	[25]
Norovirus	$10–10^3$	Acute gastroenteritis	[26]
Rotavirus	$0–10^3$	Diarrhea, gastroenteritis	[27]
Enterovirus	$10^3–10^5$ (PFU/mL)[a]	Gastrointestinal infection	[28]
Protozoa	(count/100 mL)		
Giardia	10–2500	Diarrhea, vomiting, fever	[29]
Cryptosporidium	0–100	Diarrhea, vomiting, fever	[30]

[a] A plaque-forming unit (PFU) is a measure used in virology to describe the number of virus particles capable of forming plaques per unit volume

a relatively low efficiency toward viruses [38–40]. Other disinfection technologies, such as ozone disinfection and chlorine dioxide disinfection, need to be prepared in-situ, which poses a potential hazard for chemical leakage [41]. In addition, ozone and chlorine dioxide can easily react with halogen and bromate substances in the water to generate carcinogenic DBPs [42]. Due to the strong oxidizing properties of ozone, macromolecular organic substances in wastewater can be degraded into small molecular substances that are easy for bioavailability, increasing the risk of harmful microorganism growth [43]. Membrane filtration technologies including microfiltration and ultrafiltration can effectively separate pathogenic microorganisms in water, whereas their high operation and maintenance costs and easy clogging properties hinder the large-scale application in actual situations [44–46].

Table 1.2 Technical and economic comparison of conventional wastewater disinfection methods [44]

Feature	Chemical-based disinfection			Physical-based disinfection		
	Cl_2	O_3	ClO_2	UV	Micro-filtration	Ultra-filtration
Bacterial disinfection performance	++	++	+++	+++	+++	+++
Viral disinfection performance	++	++	+++	+	+	+++
Protozoa disinfection performance	+	++	++	+++	+++	+++
Bacterial regrow	–	+	+	++	–	–
Decolorization and deodorization property	++	+++	+++	–	–	++
Toxicity of DBPs	+	++	++	–	–	–
Constant disinfection property	++	–	+	–	–	–
Operation and maintenance costs	++	+++	+++	++	+++	+++

– no impact, + low, ++ medium, +++ high

1.2 Electroporation Disinfection

1.2.1 Electroporation for Biomedical Application

Electroporation is a physical process that relies on a strong electric field to damage the outer structure of microbes (bacterial membranes and viral capsids). During the electroporation process, the permeability of the microbial outer structure increases due to the strong electric field applied [47–50]. Electroporation can be divided into reversible and irreversible electroporation. With the relatively low external electric field (10^4–10^5 V/m), the reversible electroporation holes generate on the cell membrane which can automatically recover after the electric field disappears. Based on this phenomenon, biomacromolecules such as plasmids, proteins, and DNA that cannot enter the cell through the membrane can be quickly involved in the cell during the reversible electroporation process while maintaining the cell viability [51, 52]. As a result, reversible electroporation technology is widely applied in gene transduction and cell fusion [53, 54]. On the other hand, when the external electric field strength

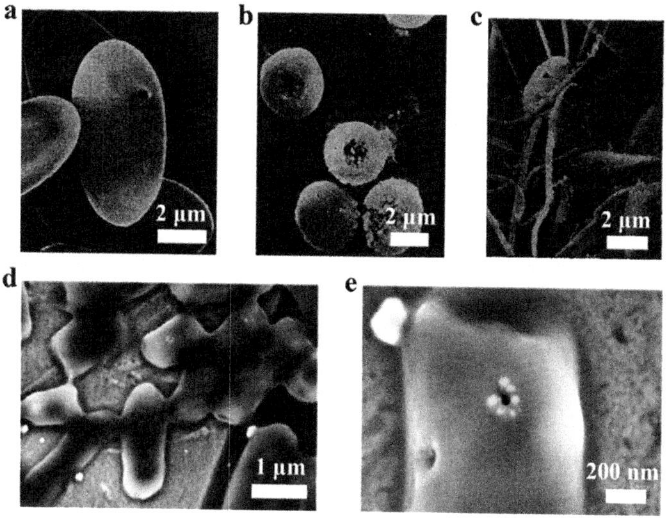

Fig. 1.1 Scanning electron microscopy (SEM) images showing cell morphologies after electroporation operation. **a**, **b** Red blood cells, **c** sperm, and **d**, **e** *Escherichia coli* (*E. coli*)

is high enough (>10^5 V/m), the electroporation hole on the cell membrane increases and expands, causing irreversible cell damage and eventually leading to cell death [55, 56]. Irreversible electroporation has the advantages of rapid and high efficiency and has been applied in cell inactivation and tumor therapy (Fig. 1.1) [57–59].

1.2.2 Electroporation for Water Disinfection

Based on the phenomenon of irreversible electroporation, electroporation microbial disinfection in water has been demonstrated with bacteria, protozoa, and viruses (Table 1.3) [60]. Compared to traditional bacteria inactivation methods (e.g., thermal, chlorine, and ultraviolet radiation), electroporation disinfection has the following advantages: (1) It is extraordinary fast (6-log inactivation can be achieved within 100 μs); (2) No chemicals are added and no disinfection byproducts are generated during treatment; (3) It is effective to all bacteria, including "superbugs" that are resistant to multiple antimicrobials; and (4) Bacteria do not develop resistance. Electroporation disinfection is also distinct from electrochemical processes where inactivation is due to in-situ generated antimicrobial chemicals (e.g., reactive oxygen and chlorine species) and/or direction oxidation on the electrode surface [61].

Current electroporation disinfection methods can achieve pathogen inactivation in water within a short contact time under specific conditions, whereas to generate a strong electric field, voltages as high as 1–10 kV are normally required, which results in high running cost and safety issues concerns. In addition, it is necessary

Table 1.3 Microbial inactivation using electroporation disinfection

Treated microbes	Energy consumption (J/mL)	Disinfection performance	References
Enterococcus faecium	120–240	>4.0 log	[62]
Pseudomonas aeruginosa	84–190	>2.2 log	[63]
Escherichia coli	80–120	>4.5 log	[64]
Listeria	40–120	>1.5 log	[64]
Saccharomyces cerevisiae	40–80	>6.0 log	[64]
Bacillus	40–80	>3.5 log	[64]
Pseudomonas putida	120	>4.0 log	[65]
Giardia[a]	~71	>2.0 log	[66]
Cryptosporidium[a]	~78	>90%	[67]

[a] The electroporation disinfection operation toward Giardia and Cryptosporidium was assisted by chlorine

to reduce the electrode spacing for a sufficient electric field. In the actual situation, small electrode spacing increases the construction footprint of the reactor and the cost of the overall facility. It is more likely that the particulate matter will be blocked between electrodes, leading to short circuits and potential safety hazards [68, 69]. Therefore, electroporation disinfection has not been successfully implemented so far and the technical limitations to achieve a high-strength electric field without the operational risk at a large scale have become the major obstacle.

1.3 Current Research Status of Novel Electroporation Disinfection

1.3.1 Nanowire-assisted Electroporation for Water Disinfection

The rapid development of nanomaterial technology provides new opportunities for solving the technical limitations in the electroporation disinfection. Nanomaterials are specific materials with dimensions (i.e., length, width, height, diameter, and/or thickness) less than 100 nm. Due to their small size, large specific surface area, and sufficient active sites, Nanomaterials are widely applied in material science, life, chemistry, and environmental fields [70, 71]. Among a variety of nanomaterials, nanowires are special materials whose length is in the micron level and the diameter is in the nanometer level [72, 73]. Due to their extremely high aspect ratio, nanowires show unique optical, electrical, and biological properties [74, 75]. When the nanowire is placed vertically on the surface of the electrode and connected to the power supply, its unique *lightning rod effect* can promote the formation of a strong electric field

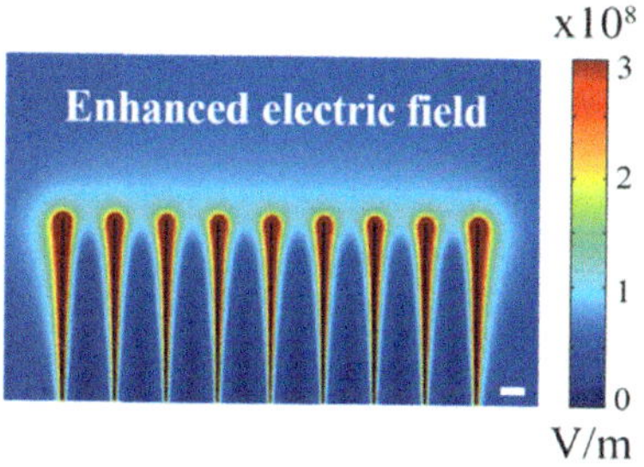

Fig. 1.2 Simulation of electric field distribution near the nanowire tip structure (diameter, 50 nm; length, 5 μm) with 1 V external voltage, showing the enhancement of the localized electric field. The scale bar is 500 nm [78]

in the tip area [76]. Even when the external voltage decreases as low as 1 V, the electric field near the tip of the nanowire can still reach 10^5 V/m, which is sufficient for irreversible electroporation (Fig. 1.2) [77].

Electroporation-based microbial disinfection has been demonstrated with bacteria, protozoa, and viruses using different nanowire-modified electrodes (Table 1.4). However, the practical application of electroporation disinfection technology is still limited due to the following problems: First, the external voltage is still high. During the electroporation disinfection process, when the DC power supply voltage is higher than 10 V, the microorganisms in the water can be efficiently inactivated in a short time. However, the continuous 10 V direct current (DC) voltage can cause a significant electrolysis water reaction, leading to unnecessary energy consumption. In addition, for wastewater containing Cl^-, chlorine is generated during the electrolysis process, which increases the risk of generation of harmful DBPs. Second, the current studies commonly test the disinfection performance in synthetic water samples, and there is a lack of systematic research on the impact of water quality of reclaimed water in actual situations for electroporation disinfection. Third, some studies integrate nanowire-assisted electroporation disinfection with chemical electrocatalysis, where the mechanism and contribution of nanowire electroporation to microbial disinfection are still unclear. Fourth, during the electrode fabrication process, some nanowires are fixed on the electrode surface by adsorption, and the adhesion is weak, which is easily detached during operation. The detachment of nanowires from electrodes not only causes the release of nanomaterials to the effluent involving potential ecological risks but also affects the lifetime of the electrode.

In summary, the primary concern for the practical application of nanowire-assisted electroporation disinfection is to optimize the reactor design and electrode preparation method for a higher disinfection efficiency and longer electrode lifetime. A reasonable reactor design can significantly increase the disinfection performance of electroporation as well as reduce the external voltage required for microbial inactivation. When the external voltage is as low as 1 V, no electrolysis reaction occurs during the disinfection process, which can significantly reduce the potential for generating harmful DBPs. In addition, a feasible nanowire-modified electrode fabrication process can ensure a higher electrode stability for a longer lifetime and a trace nanowire release.

Table 1.4 Applications using nanowire-assisted electroporation for water disinfection

Electrode materials	Power supply	Disinfection performance	Contact time	References
Ag nanowire-modified carbon fiber	DC supply 20 V	>98%	10 s	[79]
Ag nanowire-modified sponge	DC supply 10 V	>6.0 log	1 s	[77]
TiO_2 nanotube-modified electrode	DC and AC supply 2 V	>80%	15 min	[80]
CuO nanowire-modified cupper mesh	Triboelectric charge 1000 V, 1 uA	>6.0 log	1 s	[81]
ZnO nanowire-modified carbon cloth	Triboelectric nanogenerator 50 V, 1 uA	>6.0 log	10 s	[82]

1.3.2 Current Reactor for Nanowire-assisted Electroporation Disinfection

During the nanowire-assisted electroporation disinfection process, commonly, the nanowire needs to be vertically loaded on the electrode surface and an external voltage is necessary to connect with two electrodes for the power supply. When the external power supply is turned on and the water flows through the electrodes, the bacteria in the water approach the surface of the electrode, and irreversible electroporation occurs in the area where nanowires exist. Due to the enhanced electric field only existing near the nanowire tip, unless the microbes approach the tip structure, disinfection won't occur [47, 60]. Thus, the microbial transport process is the speed imitating step of nanowire-assisted electroporation disinfection. Rational reactor design can effectively accelerate the process of bacteria approaching the electrode surface. However, the studies for reactor design are limited in the current research for nanowire-assisted electroporation disinfection technology.

The reactor design can refer to the reaction device involved in the treatment of pollutants degradation in water using electrochemical processes. Common reaction devices show potential to be applied in nanowire-assisted electroporation disinfection mainly include the stirred reactor plug-flow reactor, and flow-through reactor (Fig. 1.3) [68, 83, 84]. The stirred reactor can be classified into two groups based on the operation mode of applied materials: (1) complete mixing stirred reactor with particle-sized nanomaterial complete mixed with microbes and (2) electrode stirred reactor with nanostructures immobilized or grown on panel electrodes. In a stirred reactor, rapid stirring is commonly required to drive the microbes to approach the electrode surface. Considering the mixer is necessary for the stirring process, the electrode spacing is usually more than 5 cm. This feature leads to a long traveling time for microbes to contact the electrodes, resulting in lower treatment efficiency [83]. To accelerate the rate of microbes, approach the electrode surface, researchers used a plug flow reactor where the electrode spacing can be greatly reduced, and

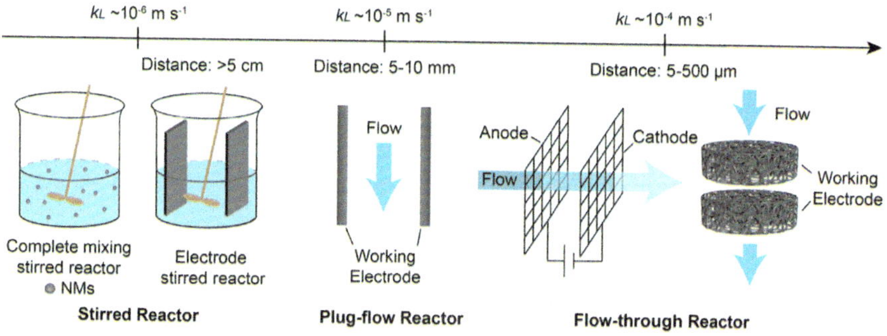

Fig. 1.3 General schematics of established nanomaterial-enabled water disinfection devices including (1) stirred reactor, (2) plug-flow reactor, and (3) flow-through reactor. Mass transfer coefficients and the distance for microbes transferring to the electrode surface in each device were summarized. k_L is the mass transfer coefficient [84]

microbes can be close to the electrode surface as the water flows through [83, 85]. However, due to the lowered electrode spacing during operation, not only will the electrode blockage leading to short circuits, but also increase the amount of electrode used, increase the construction cost, and cannot be applied on a large scale [84].

Considering that NM-enabled disinfection processes are mostly heterogeneous processes, the design of reactors with sufficient treatment rates is an impending challenge in the face of mass transfer limitations. Neither the conventional completely mixed reactor nor plug-flow reactors can provide high-efficiency and large-scale application reactors. Researchers developed a new concept using filtration inside the electrodes to achieve high-efficiency water disinfection [68, 84, 86]. Thus, flow-through devices with membrane or porous electrodes have been studied to improve treatment rates involving mesh-based electrodes or membranes including Ti mesh and CNT-membrane as well as 3-dimensional (3D) porous electrodes including carbon-based and metal foam-based porous materials [87–89]. In the flow-through disinfection device, microbial solution flow perpendicular to and through the membrane or porous electrodes with a superior mass transfer rate (~10^4 m/s) compared with those of stirred and plug-flow reactor, which overcomes mass transfer limitations [90]. For mesh materials, the electrode pore size can be as low as 10 μm and the use of porous materials as electrodes can be further reduced [59, 88]. Furthermore, attributed to the porous material with increased specific surface area of the electrode, the contact efficiency between microbes and electrode can be further improved. Meanwhile, in porous electrodes, the water flow pattern is more complicated, and microbes have more opportunities to contact with the electrode surface, which can also effectively improve the treatment efficiency [91, 92].

In summary, the flow-through reactor can be used in the nanowire-assisted electroporation process to improve the disinfection efficiency. Evenly loading the nanowires on the inner surface of the porous material allows bacteria to pass through the electrode, which can significantly increase the speed of bacteria approaching the electrode surface, improving the disinfection efficiency, and reducing the operating voltage. However, current research on the use of porous materials for the disinfection of nanowire-assisted electroporation is still limited.

1.3.3 Methods for In-Situ Nanowire Fabrication

There are two common methods for fabricating nanowires on the surface of porous electrodes. The first method is to synthesize nanowires with ideal morphology and then load the prepared nanowires on the electrode surface. The second method is to grow nanowires with ideal morphology in-situ on the electrode surface. Due to the poor stability of the loaded nanowires for the electrode prepared by the first method, nanowires cannot attach to porous electrodes tightly, which affects the stability of the electrodes [59, 77]. In addition, it is difficult to load the nanowires vertically on the electrode surface during the loading process, which affects the electric field enhancement. In contrast, the nanowire-modified porous electrode prepared by in-situ growth not only has a stronger binding force between nanowires and the electrode surface but also ensures the electrode's durability [93, 94]. The in-situ fabrication method can enable the nanowires to grow vertically on the electrode surface to achieve electric field enhancement for electroporation disinfection [95].

A porous material is one type of material with pores (cavities or channels) inside. The characteristics of porous materials depend on the size, arrangement, and shape of the pores, as well as the characteristics of through-holes (open, semi-open, or closed), and porosity (the ratio of the total pore volume to the apparent volume of the material) [96, 97]. Due to the special porous structure, the porous material has excellent features such as high specific surface area, high permeability, high porosity, and thus, has a wide range of applications in chemical, biological, pharmaceutical, and environmental remediation [98–100]. Based on the in-situ fabrication strategy, the nanowire modified porous electrode need to meet the following prerequisites: (1) It can be used as a precursor for the in-situ growth of nanowires; (2) Compared with the porous material used for adsorption, the porous electrode used for nanowire-assisted electroporation disinfection need a larger pore size (10–500 μm); (3) For a high-throughput water flux, the porous material needs a high through porosity (>99%); (4) To achieve low applied external voltage, the porous material is supposed to have a high conductivity; (5) In addition, the porous material need to be cheap, easy preparation, and high chemical and physical stable.

In summary, foam-shaped metal materials can be used as the ideal material for in-situ nanowire fabrication. Common industrially produced metal foams include aluminum foam, nickel foam, and copper foam. Among them, copper foam has a high stability and can in-situ fabricate the copper-based nanowires on the surface readily, making it an ideal material to prepare the nanowire-modified electrodes (Fig. 1.4) [101, 102]. The technology of in-situ fabrication of the copper-based nanowires on the surface of the copper foam can be generally classified into two categories: chemical vapor deposition and liquid-phase preparation method:

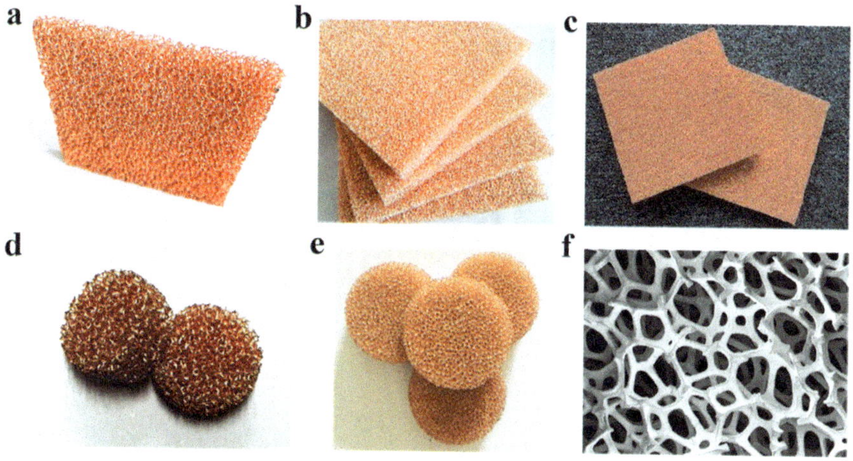

Fig. 1.4 Optical photos (**a–e**) and scanning electron microscope (SEM) images (**f**) of copper foam [91]

(1) Chemical vapor deposition

The method of chemical vapor deposition such as heating the copper substrate in air can fabricate the copper oxide (CuO) nanowires [103]. After heating the cleaned copper substrate to the aimed temperature (400–700 °C) at a specific rate followed by a constant heating at aimed temperature, CuO nanowires can grow evenly and vertically on the surface of the copper substrate (Fig. 1.5). This synthesis method is also called the thermal oxidation method. Because the thermal oxidation method is simple to operate and the synthesized nanowires are uniform and stable, it shows great potential for applications in environmental monitoring, energy, and biology [104–108].

(2) Liquid-phase preparation

Copper hydroxide ($Cu(OH)_2$) nanowires can be synthesized in-situ on the substrate surface by immersing the copper substrate in solution with special oxidation conditions. The common liquid-phase preparation methods can be classified into chemical and electrochemical liquid-phase preparation methods [109–111].

In chemical liquid-phase preparation, $Cu(OH)_2$ nanowires can be fabricated by immersing the copper substrate in an alkaline solution containing ammonium persulfate. After a certain reaction time, $Cu(OH)_2$ nanowires with ideal morphology will grow on the surface of the copper substrate (Fig. 1.6). For the electrochemical liquid-phase preparation methods, $Cu(OH)_2$ nanowires can be fabrication as follows: the copper substrate is immersed in an alkaline solution and connected to the positive electrode of the power supply. When turning on the power supply, $Cu(OH)_2$ nanowires can be synthesized on the surface of the substrate by reacting for a certain time at a specific current density (Fig. 1.7).

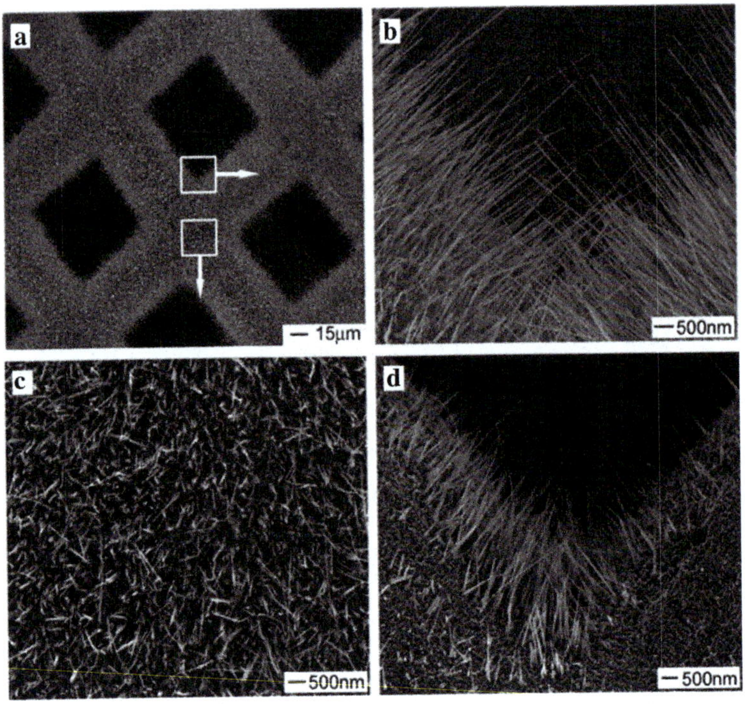

Fig. 1.5 SEM images of CuO nanowires on copper mesh [103]

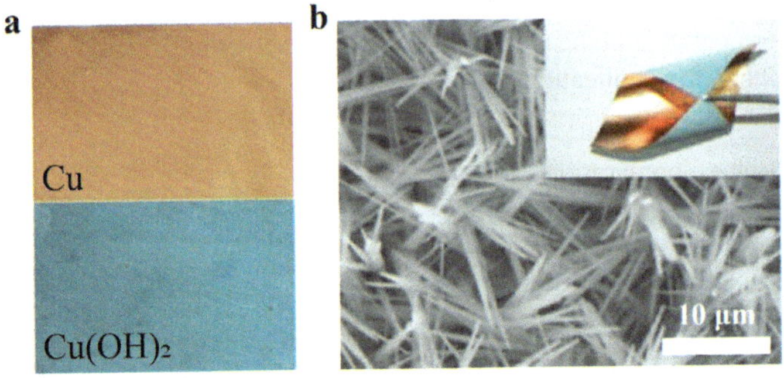

Fig. 1.6 Optical photos and SEM images of $Cu(OH)_2$ nanowires on copper foil using the chemical liquid-phase preparation methods [112]

Prepared by the above liquid-phase synthesis methods, $Cu(OH)_2$ nanowires show ideal morphology and orientation (perpendicular growth on the substrate surface). However, current research of the Cu-based nanowires only focuses on fabricating nanowires on the substrate with plat and simple structures (copper foil, copper mesh, and copper wire). Limited studies have reported the growth of $Cu(OH)_2$ nanowires

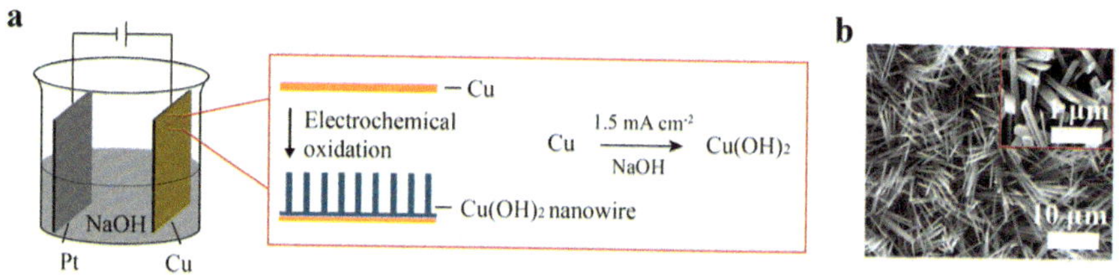

Fig. 1.7 Schematic showing the fabrication process of $Cu(OH)_2$ nanowires using the electrochemical liquid-phase preparation method and SEM images of prepared $Cu(OH)_2$ nanowires on copper foil [111]

on the surface of copper foam substrates using the liquid-phase preparation method [113]. In addition, the poor chemical stability of $Cu(OH)_2$ significantly limited the potential for practical applications, and $Cu(OH)_2$ nanowires need to be further transformed into a material with higher stability to fit application demands in the actual situations [109, 112, 114]. Phosphidation treatment is an effective method to strengthen $Cu(OH)_2$. The copper phosphide (Cu_3P) nanowires fabricated by phosphidation treatment show high electrical conductivity and chemical stability, which can be used as an ideal nanowire material suitable for electroporation disinfection. The phosphidation process is as follows: a high-temperature zone is set in the middle of the tube furnace and a low-temperature zone is set downstream of the furnace. The temperatures of the high-temperature and low-temperature zones are set at 300 and 100 °C, respectively. The prepared $Cu(OH)_2$NW-modified copper electrode is placed at the downstream side of sodium hypophosphite (NaH_2PO_2) for 90 min with Ar flushing for it to undergo a phosphidation process. After phosphidation, the $Cu(OH)_2$NWs were then converted to the Cu_3PNWs while retaining a similar morphology.

1.3.4 Impact of the Nanowire Morphology on Electroporation Disinfection

When nanowires are placed vertically on the electrode surface, the electric field on the electrode surface can be significantly enhanced, and the electric field strength can be estimated by Eqs. (1.1) and (1.2) as follows [115]:

$$\frac{E_{tip}}{E_0} = \alpha \cdot \frac{l_{rod}}{d_{rod}} \tag{1.1}$$

$$E_0 = \frac{U}{d} \tag{1.2}$$

where E_0 is the strength of the bulk electric field between two electrodes, α is a constant, l_{rod} and d_{rod} are the length and the diameter of the nanowire, U is the external voltage applied to electrodes, and d is the distance of the electrodes. The length, diameter, and aspect ratio (length: diameter) of the nanowires can significantly impact the electric field enhancement. In the current in-situ synthesis of Cu-based nanowires, researchers have pointed out that the diameter and length of CuO nanowires can be effectively controlled by changing the heating time, heating rate, and heating atmosphere components including oxygen content and humidity during the thermal oxidation process [103, 116, 117].

The diameter of CuO nanowires can be effectively adjusted by changing the heating temperature during the thermal oxidation process. It was found that the high heating temperature led to a small nanowire diameter. CuO nanowires with diameters of 100, 50, and 30 nm were obtained after treatment at 400, 500, and 600 °C (Fig. 1.8). Meanwhile, researchers also found that the nanowire length could be effectively controlled by adjusting the heating time during the thermal oxidation heating process (400 °C). After heating for 0.5, 2, and 4 h, CuO nanowires with lengths of about 1, 3, and 10 μm were obtained, respectively [105, 118, 119]. Furthermore, researchers suggested that the air humidity and oxygen content during thermal oxidation have a significant impact on both nanowire length and diameter [120].

However, it is noteworthy that changing a single experimental condition can affect both the length and diameter of CuO nanowires, and there are no systematic studies on the morphological modulation of CuO nanowires. In addition, the existing studies on CuO nanowire morphology were carried out on flat Cu foil substrates, while the study of CuO nanowire morphology on complex substrates (i.e., copper foam) is still not clear. Finally, the effect of different CuO nanowire morphologies on the electric field enhancement and the effect of electroporation disinfection has not yet been studied.

1.3.5 Nanomaterial Strengthening Method and Electrode Lifetime Improvement Method

Due to the long length (>10 μm) and small diameter (<500 nm), the nanowire material is susceptible to breakage during electroporation disinfection [121]. Meanwhile, bacteria (diameter ~1 μm) in the water will contact and hit the nanowires, further aggravating the breakage and loss of nanowires. In addition to the physical loss caused by the water flow, the electrodes need to be connected to external applied voltages during the disinfection process, which inevitably leads to electrochemical corrosion [122]. Even at low voltages (<1 V), which cannot cause water electrolysis, copper-based nanowires are still dissolved due to electrochemical reactions. Both of these reasons (physical scouring and electrochemical corrosion) will reduce the stability of the nanowire-modified electrodes and affect the electrode lifetime. This will also lead to a certain amount of copper ion release to the effluent water, which affects the treated water quality [77]. Therefore, in the nanowire-assisted electroporation disinfection process, nanowire loss needs to be minimized. First, nanowire

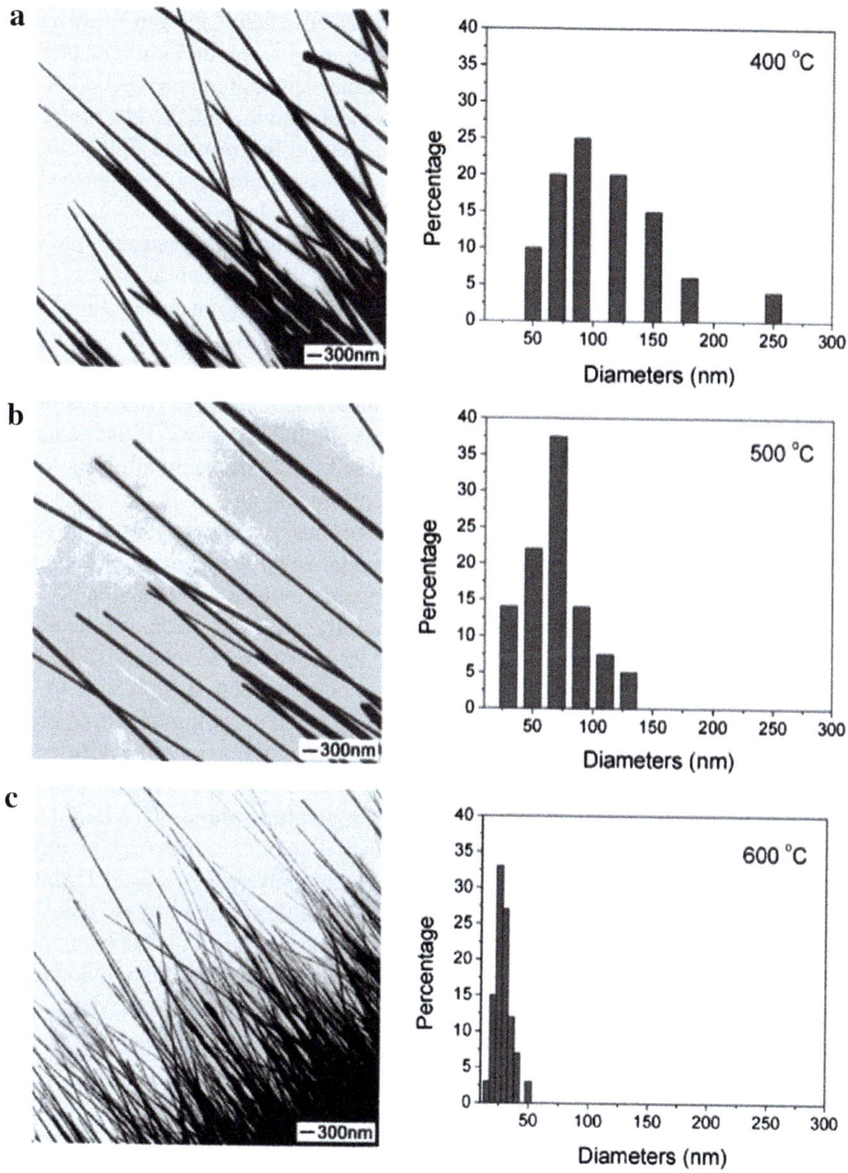

Fig. 1.8 In-situ growth of CuO nanowires on copper grids after thermal oxidation process with different heating temperatures. The corresponding nanowire diameter distribution statistics were performed according to the transmission electron microscopy (TEM) images [103]

materials with stable physical structure and good electrochemical stability are necessary. In addition, appropriate protection methods are also important to enhance the stability of nanowire-modified electrodes. Since the electrochemical reaction occurs mainly on the solution-electrode interface, the most common method of protection is to completely encapsulate the nanomaterials with a uniform and tightly loaded nanoscale inert film [123]. In nanowire-assisted electroporation disinfection, the inert nano protective film not only enhances the mechanical stability of the nanomaterials against water washout but also isolates the nanowires from water contact, which can significantly reduce electrochemical corrosion during the disinfection process [124].

Common nanosized protection materials include metallic materials (gold, titanium, aluminum), carbon materials, and organic polymers [125–128]. The electric field enhancement effect during nanowire-assisted electroporation disinfection highly relies on the aspect ratio (length: diameter) of nanowires. If a conductive coating layer is covered on the surface that exceeds the thickness of the nanowire diameter, the electric field enhancement effect will be significantly affected, which ultimately affects the disinfection effect. Therefore, organic polymers with poor electrical conductivity can be the ideal protection materials. Polydopamine (PDA) is one kind of organic polymers consisting of oxidative polymerization of dopamine molecules. Due to the unique properties of strong intermolecular interaction between PDA molecules, high plasticity, high bonding ability to the substrate, and the ability to polymerize into a thin film on the surface of arbitrary shaped materials, the PDA shows great potential to protect the nanowire-modified electrodes [129–131]. The synthesis process of PDA is simple and controllable, and the commonly used preparation method is the chemical oxidative polymerization of dopamine molecules in a solution system. The polymerization of PDA is achieved by immersing the substrate to be protected in an open container with dopamine alkaline solution (Tris buffer solution, $pH = 8.5$), which allows the polymerization of dopamine molecules with oxygen in the air [131, 132]. PDA films can be synthesized not only on the surface of arbitrary solid materials but also in larger thicknesses by multiple coating cycles [129, 133]. The polymerization process of PDA deposition on the surface of nanowire-modified electrodes is shown in Fig. 1.9. However, the existing studies on the polydopamine protection of complex substrates (e.g. copper foam) are still limited. Meanwhile, the effect of PDA films on the disinfection performance for the nanowire-assisted electroporation disinfection is not clear.

In addition to applying thin films to protect nanowire-modified electrodes to enhance stability, optimization of reactor operating conditions can also effectively enhance electrode lifetime. In the nanowire-assisted electroporation disinfection process, an external voltage is necessary. When the external voltage is lower than 1 V, it can effectively reduce the electrolytic water reaction, but it still inevitably causes electrochemical corrosion of the copper-based electrode. At this point, a high-frequency altering current (AC) power supply can be used instead of a direct current (DC) power supply to reduce or eliminate electrode loss caused by electrochemical corrosion. During the treatment process of applying voltage to the electrode, there are different physical and chemical processes, including electrochemistry and electroporation. Electrochemistry has its own specific response time, usually in the range

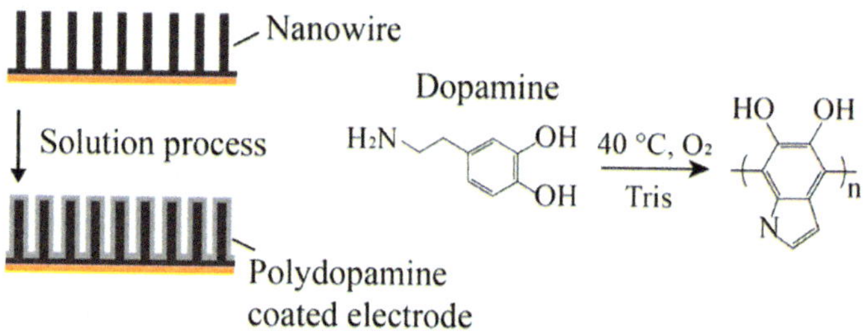

Fig. 1.9 Polymerization process of PDA deposition on the surface of nanowire-modified electrodes

of ms level. If the period of AC applied to both electrodes is less than ms (i.e., the frequency is greater than 10^3 Hz), the electrochemical reaction in the solution can be weakened [134, 135]. While, for the electroporation process, it usually takes a few μs to generate transmembrane voltages on both sides of the cell membrane and to achieve electroporation [136, 137]. Therefore, electroporation has its own specific response time (μs level), and if the applied AC period is less than μs (i.e., the frequency is greater than 10^6 Hz) electroporation will not occur (Fig. 1.10). Based on this principle, electroporation disinfection can theoretically be achieved while shielding the electrochemical reaction by controlling the AC frequency from 10^3 to 10^6 Hz (Fig. 1.10). By this method, electrochemical corrosion of the electrodes can be completely eliminated to maximize their service life. However, the current research is still limited.

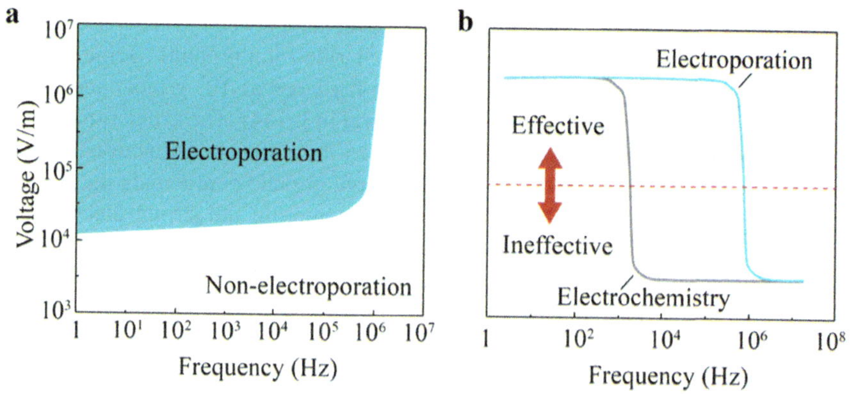

Fig. 1.10 **a** Schematic showing the impact of AC frequency and electric field strength on electroporation, **b** schematic showing the effect of AC frequency on electrochemistry and electroporation

1.3.6 Treatment Efficiency of Nanomaterial-Enabled Disinfection Technology for Reclaimed Wastewater

During the actual treatment process, the disinfection performance of nanowire-assisted electroporation may be influenced by water quality parameters (e.g., ionic strength, ion type, pH, and organic concentration and type). The core process during nanowire-assisted electroporation disinfection is the generation of voltage differences inside and outside the cell membrane using a strong external electric field, which eventually leads to the occurrence of electroporation. However, during electroporation, the ions in the solution may react electrochemically due to the contact between the electrode and the solution [138]. Therefore, the ionic strength and ionic species of the solution may impact the disinfection, but there are no relevant studies yet. pH may also be another important factor affecting nanowire-assisted electroporation disinfection [139]. The isoelectric point of bacteria in solution is usually between pH 3 and 5 (pH = 3–4 for Gram-positive bacteria; pH = 4–5 for Gram-negative bacteria) [140–142]. Normally, the natural water pH is neutral (pH = 7). Thus, the bacteria existing in that water are commonly negatively charged and will be attracted to the positive electrode and repelled by the negative electrode. However, as the pH changes, the surface charges of bacteria charge leading to the change of the gravitational and repulsive forces by the electrodes, which finally impact the disinfection efficiency. While the relevant studies are still limited.

During the practical application of the nanowire-assisted electroporation disinfection process, another important water quality parameter that may affect disinfection performance is the type and concentration of organic matter. There is a certain concentration of macromolecular organic matter in the reclaimed water, usually secreted by microorganisms in the water, mainly polysaccharides and proteins. Some of these macromolecules have large molecular weight (as large as 10^5 g/mol) and have a non-negligible spatial size (several hundred nm) [143, 144]. These macromolecular organics tend to adhere to solid surfaces due to their abundant functional groups [145, 146]. Therefore, it is possible that the large organic compounds may adhere to the nanowire tips and prevent bacteria from entering the strong electric field region enhanced by the nanowires, which ultimately affects the disinfection efficiency. However, the research on the effect of such organic matter on disinfection is still limited. In summary, reclaimed water is complex, containing high ionic strength and different types of ions, large pH fluctuations, high organic concentrations, and certain concentrations of macromolecular organics. It is important to study the effect of the impact of the water quality parameters on nanowire-assisted electroporation disinfection, whereas the study is still in the initial stage and lacks in-depth and systematic investigation.

1.4 Research Topics to Be Further Investigated

Disinfection is a key process in ensuring water safety and controlling biological risks during wastewater reclamation. However, due to the complex water quality conditions of reclaimed water, the commonly used chlorine disinfection technology inevitably generates disinfection by-products during the treatment process, while other disinfection technologies such as UV disinfection, ozone disinfection, and membrane filtration technologies have their own drawbacks. Biological risk control in reclaimed water reuse has become a key research issue. Nanowire-assisted electroporation disinfection technology is a new, rapid, and energy-efficient disinfection technology with great potential for industrial application. However, research on nanowire-assisted electroporation disinfection is severely limited. Further research on microbial removal characteristics, reactor optimization design, and electrode stability enhancement need to be considered and performed, which are summarized as follows:

(1) Nanowire-assisted electroporation disinfection requires bacteria to approach the electrode surface, so a reasonable reactor design will shorten the bacterial traveling time and significantly improve the disinfection efficiency. However, the research on new and efficient reactors for nanowire-assisted electroporation disinfection is still limited.
(2) The microbial removal efficiency, microbial inactivation mechanism, microbial reactivation/regrowth after treatment, and the influence of reclaimed water quality parameters during nanowire-assisted electroporation disinfection are not clear.
(3) Nanowire-assisted electroporation disinfection requires the special nanowire structure to enhance the electric field to achieve electroporation, whereas there is no research to investigate the effect of nanowire morphology on electric field enhancement and disinfection efficiency.
(4) The major challenge of limiting the practical application of nanowire-assisted electroporation disinfection is that nanowires are susceptible to breakage and electrochemical corrosion by water flow. However, there is no research on the optimization of nanowire material selection and electrode stability enhancement.

1.5 Research Objective and Content

1.5.1 Research Objective

To guarantee the microbial safety of reclaimed water, we developed a flow-through nanowire-assisted electroporation disinfection device to achieve high-performance microbial inactivation with low voltage. We aimed to clarify the electroporation disinfection mechanism, master the electrode optimization method, and clarify the surface strengthening method to achieve efficient, stable, and safe reclaimed water disinfection.

1.5.2 Research Content

The main contents of this study are as follows:

(1) Nanowire-modified electrode preparation and its microbial inactivation performance investigation.
Build a nanowire-assisted electroporation disinfection device based on the flow-through mode of porous electrodes. Study the microbial inactivation performance of nanowire-assisted electroporation disinfection devices in the reclaimed water. Understand the risk of harmful disinfection by-products during the treatment process. Investigate the mechanism of nanowire-assisted electroporation disinfection.
(2) Effect of nanowire-modified electrode structure on microbial inactivation.
Systematically study the effect of nanowire shape, electrode design parameters, and reactor design on the disinfection effect of nanowire-assisted electroporation. Clarify the effect of electrode morphology on the nanowire-assisted electroporation disinfection. Understand the effect of electrode pore size and thickness on the disinfection effect. Clarify the contribution of positive and negative electrodes to disinfection and optimize the reactor design.
(3) Preparation of high-strength nanowire-modified electrodes and their microbial inactivation performance.
Use electrochemical oxidation and phosphidation process to prepare Cu_3P nanowires with high physical strength and electrochemical stability, replacing the original CuO nanowires. Investigate the microbial inactivation performance of Cu_3P nanowires-modified electrodes. Analyze the lifetime of Cu_3P nanowires-modified electrodes. Understand the degree of electrode loss as well as loss mechanism in the long-term operation process.
(4) Surface coating on nanowire-modified electrodes to improve the electrode lifetime.
The PDA film is uniformly deposited on the nanowire-modified electrode surface by the solution method. Understand the inactivation performance and mechanisms of the nanowire-modified electrodes coated with thin PDA film. Clarify the lifetime of the electrode. Understand the degree of electrode loss as well as loss mechanism in the long-term operation process.
(5) The improvement of the lifetime of nanowire-modified electrode driven by AC power supply.
The use of an AC power supply instead of the original DC power supply mode ensures the effect of electroporation disinfection while preventing the occurrence of electrochemical reactions. Understand the inactivation performance

and mechanisms of nanowire-assisted electroporation disinfection with AC power supply. Investigate the electrode lifetime under such operating conditions. Clarify the degree of electrode loss as well as loss mechanism in the long-term operation process.

(6) Performance of nanowire-assisted electroporation disinfection for reclaimed water.
Systematically analyze the effect of reclaimed water quality conditions on nanowire-assisted electroporation disinfection. Understand the disinfection efficiency of nanowire-assisted electroporation disinfection technology under reclaimed water quality conditions. Clarify the effect of nanowire-assisted electroporation disinfection on actual reclaimed water disinfection.

1.5.3 Research Roadmap

The research roadmap of the thesis is shown in Fig. 1.10. Firstly, a nanowire-modified electroporation disinfection device based on the flow-through treatment mode of porous electrodes was constructed, and the device was used to analyze the disinfection of typical model microorganisms in reclaimed water. The microbial reactivation/regrowth phenomenon during the storage process after disinfection was also investigated. On the basis of this study, electrode optimization was carried out to analyze the influence of nanowire morphology and electrode parameters on disinfection efficiency. High-strength nanowires were prepared and surface strengthening technology as well as high-frequency AC power supply mode were used to enhance electrode stability. Finally, the disinfection efficiency of reclaimed water by nanowire-assisted electroporation disinfection was investigated. Our work further clarified the effect of water quality conditions on the disinfection of reclaimed water and investigated the disinfection effect of nanowire-assisted electroporation on actual reclaimed water. In summary, this thesis provides theoretical and technical support for nanowire-assisted electroporation disinfection to ensure the biosafety of wastewater reclamation and reuse (Fig. 1.11).

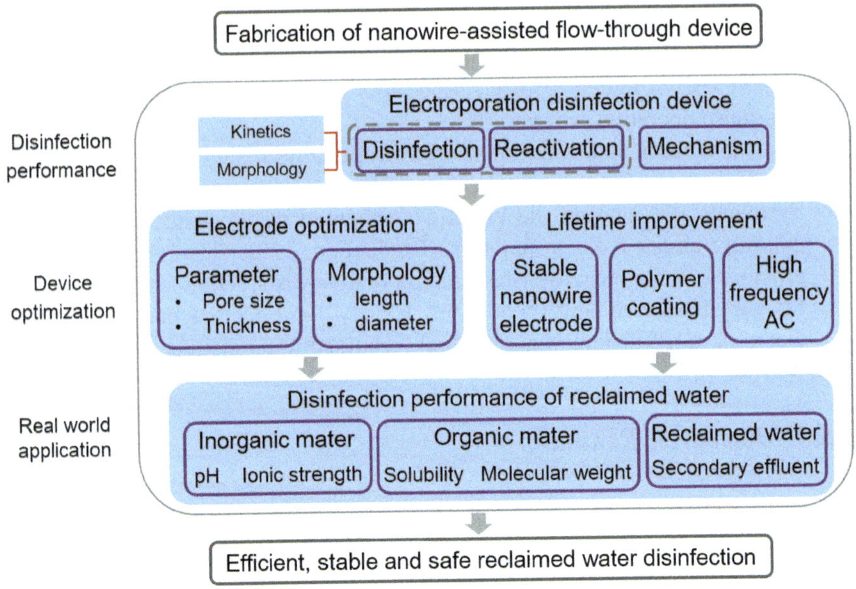

Fig. 1.11 Research roadmap

Chapter 2
Development of Nanowire-Modified Electrodes and Investigation of the Microbial Inactivation Performance

2.1 Research Background

Bacteria can undergo irreversible electroporation reactions in the region of the nanowire-enhanced electric field, leading to cell death [57, 147]. Nanowire-assisted electroporation disinfection is a physical process with the advantages of high efficiency, low energy consumption, no chemical dosage, and no harmful disinfection by-product generation, which is expected to be used for microbial disinfection treatment during wastewater reclamation and reuse process [60]. In the nanowire-assisted electroporation disinfection process, bacteria are supposed to be close enough to the electrode surface with a nanowire-enhanced electric field. Therefore, the bacteria approach the electrode surface becomes the speed-limiting process for nanowire-assisted electroporation disinfection. Through efficient reactor design, the use of porous electrodes instead of existing flat or mesh electrodes can effectively increase the rate of bacteria traveling to the electrode surface and enhance the disinfection efficiency [52]. However, at present, the uniform growth of nanowires with ideal morphology on the surface of porous electrodes is still questionable, and the relevant studies are severely limited.

The objectives of this chapter are to fabricate the nanowires with ideal morphology that can grow uniformly on the surface of porous electrodes. A new flow-through electroporation disinfection device that allows water to pass through the inner part of the electrode will be built using the prepared nanowire-modified electrodes. The established disinfection device will be used to evaluate the disinfection performance of typical model microbes in reclaimed water. The energy consumption and the phenomenon of disinfection by-product generation will be investigated during the treatment process. In addition, the microbial disinfection mechanism will be analyzed and the changes in bacterial concentration during the storage process after disinfection will be examined.

Z. Huo, *Nanowire-assisted Flow-through Electrode Enabling Electroporation Disinfection of Reclaimed Water*, Springer Theses,
https://doi.org/10.1007/978-981-99-4502-3_2

2.2 Experimental Materials and Methods

2.2.1 Experimental Reagents

Hydrochloric acid (HCl), nitric acid (HNO_3), anhydrous ethanol, glycerol, sodium chloride, phosphate buffer solution (PBS; 1 mol/L, pH = 7.4), glutaraldehyde, and tert-butanol used in this study were purchased from Beijing Bailingway Technology Co. Live/Dead Baclight kits for bacterial staining were purchased from Thermo Fisher, U.S.A. Luria–Bertani (LB) liquid medium, LB agar medium, nutrient broth liquid medium, and nutrient agar medium were purchased from Shanghai Beinuo Biotechnology Co. Deionized (DI) water was prepared by a laboratory water purification system (Millipore, USA). The solutions used in the experiment were prepared from deionized water.

2.2.2 CuO Nanowire-Modified Copper Foam Electrodes Fabrication and Disinfection Device Construction

Copper foam (Jiayisheng Electronics Co., Ltd. in Kunshan, China) was selected as the porous electrode material and fabricated into a cylindrical electrode with a diameter of 10 mm and a thickness of 5 mm using laser cutting. The pore size of the copper foam electrode was 500 μm and the electrode porosity was greater than 95%. The CuO nanowire-modified copper foam electrode was prepared by the thermal oxidation method. Firstly, the copper foam was immersed in 1 mol/L HCl solution for 1 min to remove the oxide layer on the surface of the copper foam electrode. After that, it was rinsed with DI water for 30 s to wash off the residual solution on the surface of copper foam, and then repeatedly washed three times. After the final washing, the electrode was dried using N_2 gas. Finally, the prepared copper foam was heated to 400 °C at a heating rate of 3.3 °C/min and kept at 400 °C for 2 h using a tube furnace (Tianjin Zhonghuan Electric Furnace Co., Ltd., SK-G06123K). The electrode synthesis process is shown in Fig. 2.1. After heating, the CuO nanowire-modified copper foam electrode was prepared, and the electrode color changed from brown to black.

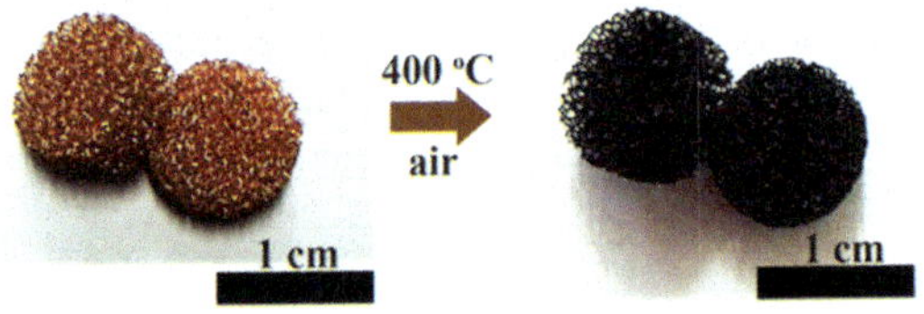

Fig. 2.1 Fabrication of CuO nanowire-modified copper foam electrode

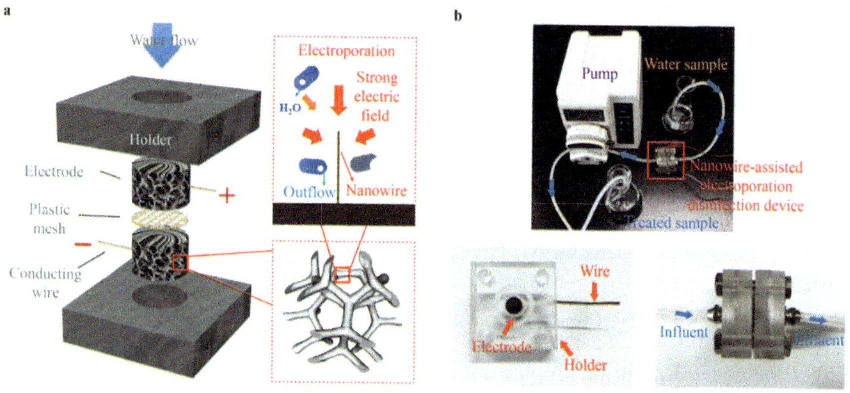

Fig. 2.2 **a** Schematic showing the nanowire-assisted electroporation disinfection device with CuO nanowire-modified copper foam electrode. **b** Optical photos of the experiment setup during operation. Bacteria in water samples pumped by peristalsis pump flew through the nanowire-assisted electroporation disinfection device and were inactivated by electroporation disinfection

The prepared CuO nanowire-modified copper foam electrode was fixed in a Plexiglas electrode holder, and the electrodes were separated by a plastic mesh (200 μm pore size) to prevent electrode contact and short circuit. The two electrodes were connected to the positive and negative DC power supply (Beijing Dahua, DH1715A) through wires, respectively (Fig. 2.2a). In the actual treatment process, a peristaltic pump was used to drive the water sample to pass through the CuO nanowire-modified copper foam electrode at the specified flow rate (Fig. 2.2b). At the same time, the power is turned on and a specific external voltage is applied between the two electrodes. As the bacteria flow through the electrode, electroporation disinfection occurs in the area of the electrode surface with the nanowire-enhanced electric field.

2.2.3 Characterization of CuO Nanowire-Modified Copper Foam Electrodes

Morphological analysis of CuO nanowires-modified with copper foam electrodes was performed using an SEM (Hitachi, SU8220) at an operating voltage of 5 kV. Morphological analysis and structural characterization of single CuO nanowires were carried out using TEM (JEOL, JEM-200CX) at an operating voltage of 160 kV. The TEM samples were prepared as follows: (1) firstly, the CuO nanowire-modified foam copper electrode was sonicated in a solution containing anhydrous ethanol for 5 min to detach the CuO nanowires from the electrode surface; (2) after sonication, the suspension containing the CuO nanowires was placed on a TEM grid (100 mesh, CEMCO); (3) the prepared grids were dried in an oven at 60 °C for 20 min to remove the ethanol solution from the sample surface.

2.2.4 Microbes and Water Samples Used in Experiments

(1) Microbes used in experiments

The strains used in this study were acquired from the China General Microbiological Culture Collection Center (CGMCC) including *Escherichia coli* (*E. coli*; CGMCC 1.3373), *Enterococcus faecalis* (*E. faecalis;* CGMCC 1.15424), and *Bacillus subtilis* (*B. subtilis*; CGMCC 1.821). All bacteria were stored in a refrigerator at −80 °C.

(2) Water samples used in experiments

E. coli suspension solution preparation: Transfer 10 μg of *E. coli* bacterial dry powder into 20 mL of sterile LB liquid medium (see Table 2.1 for composition) under aseptic conditions and incubate the bacteria at 37 °C, 150 rpm in an incubator with shaking for 12 h. After incubation, take 0.5 mL of the bacterial solution and add 0.5 mL of sterile glycerol solution and store it in the refrigerator at 4 °C as the bacterial stock solution.

To prepare the *E. coli* suspension for the experiment, add 1 mL of the above-mentioned bacterial stock solution to 10 mL of sterile LB liquid medium. Incubate at 37 °C, 150 rpm in an incubator for 12 h. Centrifuge the bacterial suspension at 14,500 rpm for 5 min at room temperature (Hitachi, RX2 series). Discard the supernatant and resuspend it with DI water. Repeatedly centrifuge and wash the sample for 3 times to fully wash the bacteria and remove the residual liquid medium and dilute the bacteria into deionized water to control the concentration of *E. coli* in the bacterial suspension to ~10^7 colony-forming unit/mL (CFU/mL).

E. faecalis suspension solution preparation: The cultivation steps are the same as *E. coli* suspension except that the liquid medium for culture is nutrient broth medium (see Table 2.1 for composition).

B. subtilis suspension solution preparation: the cultivation steps are the same as *E. coli* suspension.

Secondary effluent samples from municipal wastewater treatment plants: The water samples were collected from the secondary effluent of the Beijing municipal wastewater treatment plant (without disinfection), and the municipal wastewater

Table 2.1 Composition of the media used in the study

Name	Composition	Conc. (g/L)	Name	Composition	Conc. (g/L)
LB liquid medium	Tryptone	10.0	Nutrient broth liquid medium	Beef extract	3.0
	Yeast extract	5.0		Peptone	5.0
	NaCl	10.0		NaCl	10.0
LB agar medium	Tryptone	10.0	Nutrient agar medium	Beef extract	3.0
	Yeast extract	5.0		Peptone	5.0
	NaCl	10.0		NaCl	10.0
	Agar	15.0		Agar	15.0

Table 2.2 Basic water quality parameters of the secondary effluents from municipal wastewater treatment plants

Name	pH	Total dissolved solids (mg/L)	Heterotrophic plate count (CFU/mL)
wastewater treatment plants (WWTP) 1	6.8	418	12,000
wastewater treatment plants (WWTP) 2	7.2	545	35,000

treatment plants were selected attributed to their stable water volume, large scale (>10,000 t/day), and continuous operation to ensure the representativeness of the collected water samples. The actual water samples were collected in sterile containers after sampling and placed in iceboxes to ensure low temperature and protection from light. The samples were transported to the laboratory as soon as possible after sampling and the experimental tests were completed within 4 h. The basic water quality parameters of the secondary effluents from municipal wastewater treatment plants are shown in Table 2.2.

2.2.5 *Nanowire-assisted Electroporation for Microbial Disinfection*

(1) Hydraulic retention time (HRT) calculation

The volume of the reactor equals the apparent volume of two porous electrodes, which can be calculated by Eq. (2.1).

$$\text{Volume of the reactor} = 2 \times \text{ electrode bottom area } \times \text{ electrode thickness} \tag{2.1}$$

where the electrode diameter is 1 cm, and the electrode thickness is 0.5 cm. After calculation, the volume of the reactor is 162 cm^3.

The HRT in the disinfection process can be calculated by Eq. (2.2) as follows:

$$\text{HRT} = \text{Volume of the reactor/flow rate} \tag{2.2}$$

where the flow rate is controlled by the peristaltic pump.

(2) Nanowire-assisted electroporation disinfection experiment

Based on Eq. (2.2), the flow rate required for different HRT can be calculated. Control the water sample to pass through the reactor at the required flow rate using the pump. After the water sample has passed through the reactor for 1 min, turn on the DC power supply with aimed voltage. After applying the voltage for 5 min, 0.5 mL of sample was taken from the reactor effluent for further bacterial culture counting. The same experiment was repeated 3 times.

(3) Bacteria counting method

Bacteria were counted by the plate counting method. *E. coli s*uspension was diluted by sterile saline (0.9% NaCl solution) in a tenfold gradient. After gradient dilution, 100 μL bacterial solution was taken and added to LB agar medium which was poured and solidified in advance, and the liquid was evenly spread on the surface of LB agar medium using a sterile L-shape spreader. Then put the agar plate in the incubator at 37 °C incubation for 12 h. The final solid medium with 10–300 colonies counted as valid values. Each bacterial counting experiment was repeated 3 times and take the average value of the count.

E. faecalis and *B. subtilis* suspensions in the bacterial count methods are the same as the method used for *E. coli* counting except for the used medium (nutrient agar medium).

The number of heterotrophic bacteria in the actual water sample is determined using the heterotrophic bacteria plate count (HPC) method which is the same as the method counting *E. faecalis* and *B. subtilis* suspensions.

(4) Evaluation method of nanowire-assisted electroporation disinfection

In this study, the disinfection efficiency of nanowire-assisted electroporation was evaluated using the logarithmic inactivation rate based on the Eq. (2.3) as follows:

$$\text{Log-removal rate} = -\log 10(N/N_0) \tag{2.3}$$

where N is the number of live bacteria (CFU) in the water sample after nanowire-assisted electroporation disinfection by and N_0 is the number of live bacteria (CFU) in the water sample before disinfection.

For model bacterial suspensions, the bacterial concentration in the influent was controlled to be 10^7 CFU/mL. If no colonies were detected on the agar plate after disinfection, the concentration of live bacteria in the effluent water was less than 1 CFU/100 μL (i.e., 10 CFU/mL), so the log-removal rate was calculated to be >6.0 at this time.

2.2.6 Bacterial Storage After Nanowire-assisted Electroporation Disinfection

E. coli was selected as the experimental strain for all the bacterial storage experiments involved in this study. The supernatant was discarded and resuspended with DI water, saline, and sterile secondary effluent filtered by filter membrane (Millipore, USA, pore size 0.22 μm), and the resuspended bacterial solution was stored in sterile glass tubes. The bacteria were stored in sterile glass tubes and placed in a constant temperature incubator at 25 °C for 2, 4, 8, 12, and 24 h, and 100 μL of bacteria were taken under a sterile environment for plate counting. The storage experiments were repeated 3 times for each disinfectant dose.

2.2.7 Free Chlorine Detection and Current Detection During Nanowire-assisted Electroporation Disinfection

During the disinfection of different water samples, free chlorine in the effluent was detected using a free chlorine detector (Hana, HI96724, USA). The current during treatment is detected by a sensitive multimeter connected in series in the circuit (Keithley, USA, 2000).

2.2.8 Copper Ion Concentration Detection

After the reaction unit was operated for 10 min, 5 mL of effluent was collected and 5 mL of HNO_3 (2 mol/L) was added to dissolve the copper or CuO particles in the effluent. After filtering through a filter membrane (Millipore, USA, pore size 0.22 μm), the copper ion concentration was detected using inductively coupled plasma mass spectrometry (ICP-MS, Thermo Scientific, XSERIES 2).

2.2.9 Bacterial Morphology Analysis

The bacterial morphology analysis involved in this study was used *E. coli* as the experimental strain. The cell morphology was analyzed using SEM and the samples were prepared as follows:

(1) *E. coli* suspension (20 mL) was taken before and after electroporation disinfection or after different storage times and centrifuged at 14,500 rpm for 10 min at room temperature (Hitachi, RX2 series). Resuspended the sediments and fixed them overnight in the fixative containing 0.1 M PBS (pH 7.4) and 2% glutaraldehyde at 4 °C, then washed with the same PBS buffer for 5 min.
(2) The samples were dehydrated in increasing concentrations of ethanol solution (50, 70, 90, and 100%), and critical point dried in 100% ethanol with liquid CO_2.
(3) The samples were finally sputter-coated with 10 nm of gold. All the SEM images were taken by an SEM (Hitachi, SU8220) at an operating voltage of 5 kV.

2.2.10 *Bacterial Staining Experiments*

The bacterial staining experiments involved in this study was used *E. coli* as the experimental strain, and the bacterial staining experiments were conducted as follows:

(1) *E. coli* suspension (2 mL) was taken before and after electroporation disinfection or after different storage times and centrifuged at 14,500 rpm for 10 min at room temperature (Hitachi, RX2 series). The supernatant was discarded and resuspended with 0.2 mL of PBS (0.1 M; pH 7.4).
(2) Bacterial staining was performed using the Live/Dead Baclight (Thermo Fisher) kit and 5 μL SYTO9 and 5 μL Propidium Iodide (PI) reagents were added to the *E. coli* samples. A control experiment without *E. coli* was set up for staining treatment.
(3) The samples were stored at room temperature under shade for 30 min and then observed under a fluorescence microscope (DM6000B, Leica).

2.3 Fabrication of CuO Nanowire-Modified Copper Foam Electrodes

2.3.1 *Characterization of CuO Nanowire-Modified Copper Foam Electrodes*

CuO nanowire-modified copper foam electrodes were synthesized by the thermal oxidation method. The electrode structure and nanowire morphology were characterized as shown in Fig. 2.3. The copper foam electrode has a continuous open-hole structure with a pore size of about 500 μm, and the CuO nanowires grow vertically on the surface of the copper foam substrate during the thermal oxidation treatment. The length of the nanowires is about 10 μm and the diameter is about 30 nm. The synthesized nanowire material has a specific 2.32 Å crystal plane spacing according to the high-resolution TEM measurement, which matches with the specific crystal plane spacing of CuO. In addition, the electron diffraction pattern of the nanowire structure also confirmed that the monoclinic crystal of the nanowire material is CuO.

The use of CuO nanowire-modified copper foam electrode for electroporation disinfection has the following advantages: (1) Since bacteria need to be close to the nanowire-enhanced electric field area near the electrode surface for electroporation, the use of porous electrodes for disinfection by flow-through treatment can effectively increase the opportunities of bacteria approaching the electrode surface, thus improving the disinfection efficiency; (2) The large pore size copper foam electrode (pore size 500 μm) has a continuous open-hole structure, which allows the water sample to flow rapidly through the electrode and effectively reduces the washout of the electrode by the water flow; (3) The large pore size of the copper foam electrode can effectively avoid the blockage of the electrode by suspended substances

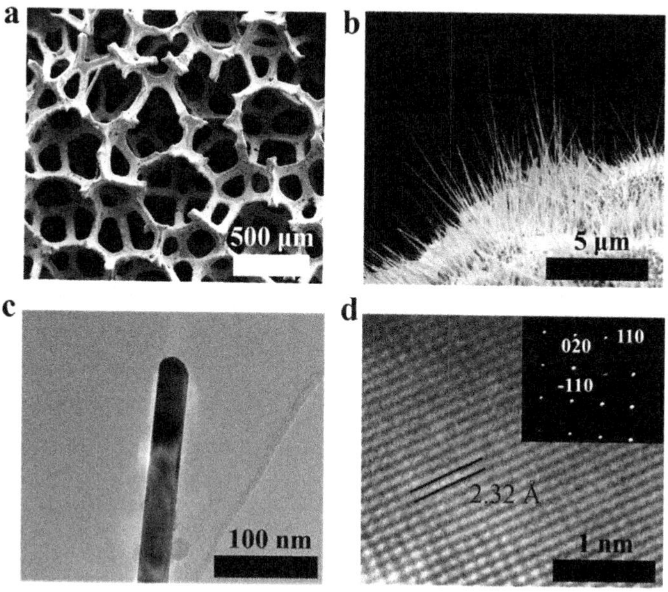

Fig. 2.3 Characterization of CuO nanowire-modified copper foam electrodes. **a** SEM image of a CuO nanowire-modified copper foam electrode showing the macroporous structure. **b** High magnification SEM image of the copper foam surface with CuO nanowires. **c** TEM image of the tip of a CuO nanowire. **d** High-resolution TEM image of the CuO nanowire and the electron diffraction pattern

in the water when treating the actual water sample; (4) In addition, because the CuO nanowires are grown *in-situ* on the surface of the copper foam substrate, the contact between the nanowires and the substrate is relatively strong, which ensures the stability of the electrode.

2.4 Disinfection Efficiency of CuO Nanowire-Modified Copper Foam Electrodes

2.4.1 Disinfection Efficiency of E. Coli

The disinfection efficiency of E. coli by CuO nanowire-modified copper foam electrodes was shown in Fig. 2.4. As the applied external voltage increases, the electroporation disinfection efficiency increased at different HRT. When the HRT was 1 s, the nanowire-assisted electroporation disinfection enabled complete inactivation of E. coli at an external voltage of 5 V, and the bacterial removal rate was >6.0 log. An appropriate increase in HRT significantly reduced the voltage required for complete bacterial inactivation. At HRT of 2 and 5 s, the external voltage required

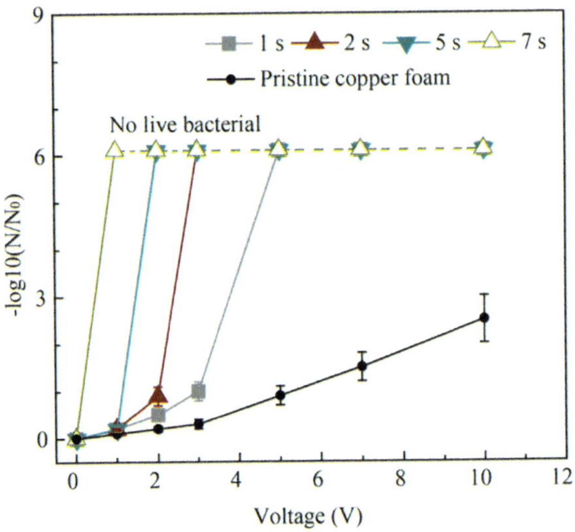

Fig. 2.4 Disinfection efficiency of CuO nanowire-assisted electroporation to *E. coli*. Dashed lines indicate that all microbes are inactivated, and no live microbes can be detected

for complete bacterial inactivation was 3 and 2 V, respectively. When the HRT was further reduced to 7 s, only 1 V external voltage was required to achieve complete bacterial inactivation.

To ensure that bacterial inactivation was only contributed by electroporation, the following control experiments were designed to exclude other potential disinfection mechanisms.

(1) Analysis of the *E. coli* removal efficiency using the disinfection device without external voltage. The results showed that without external voltage, the concentration of *E. coli* in the effluent was the same as that in the influent. This indicated that the nanowire structure had no significant effect on the cell activity without electricity, and there was no significant interception of bacteria by the electrode.
(2) Analysis of the *E. coli* removal efficiency using the copper foam electrode without CuO nanowire. The copper foam electrode without nanowires had no significant effect on the removal of E. coli at less than 3 V at an HRT of 7 s. While, under the same operating conditions, the CuO nanowire-modified electrode enabled completely bacterial inactivation. This indicated that it is the nanowire structure on copper foam that achieves efficient bacterial inactivation when applied with an external voltage.
(3) Detection of the concentration of copper ions in the effluent. Seeded *E. coli* in the LB agar medium containing copper ion. The copper ion concentration is the same as that detected in the effluent. Incubate the *E. coli* for 24 h, and then determine the number of colonies on the plate. After the treatment with HRT of 7 s, the copper ion concentration in the effluent was 320 μg/L, and the colony number of *E. coli* in the agar plates with copper ion (320 μg/L) was the same as that in the medium without the addition of copper ion after 24 h of

incubation. This indicates that the copper ions in the effluent did not contribute to the inactivation of *E. coli*.

In summary, the nanowire-enhanced electric field was the main mechanism of bacterial inactivation during electroporation disinfection. There was no significant bacterial interception during the treatment process. In addition, the contact between nanowires and bacteria and the concentration of copper ions in the water did not contribute significantly to bacterial disinfection.

2.4.2 Disinfection Efficiency of E. Faecalis, B. Subtilis, and Secondary Effluent from Municipal Wastewater Treatment Plants

The electroporation disinfection method using CuO nanowire-modified copper foam electrodes can achieve complete inactivation of *E. coli* at a short operating time (HRT = 7 s) at an external voltage as low as 1 V. Considering that *E. coli* is a Gram-negative bacterium, typical Gram-positive bacteria (*E. faecalis* and *B. subtilis*) in reclaimed water were also selected to investigate the inactivation performance of nanowire-assisted electroporation at a low voltage (1 V).

The disinfection efficiency of *E. faecalis* and *B. subtilis* by nanowire-assisted electroporation was shown in Fig. 2.5. The disinfection efficiency of nanowire-assisted electroporation for both *E. faecalis* and *B. subtilis* increased with the increase of HRT, and complete bacterial inactivation (removal rate > 6.0 log) was achieved at an HRT of 10 s. The contact time (i.e., HRT) required for complete inactivation of *E. faecalis* and *B. subtilis* was longer (10 s) than that of *E. coli*. This lowered disinfection performance is likely due to the thicker layer of peptidoglycan on the cell membrane of the Gram-positive bacteria [148].

The study further analyzed the inactivation efficiency of nanowire-assisted electroporation to the microbes in real water samples. As shown in Fig. 2.5, the heterotrophic microbes were completely inactivated in the secondary effluent of the municipal wastewater treatment plant after nanowire-assisted electroporation with an external voltage of 1 V and an HRT of 10 s. No live bacteria were detected in the treated effluent. The above results indicate that the nanowire-assisted electroporation disinfection technique with CuO nanowire-modified copper foam electrodes can achieve efficient inactivation of pure bacteria and microbes in real water samples. The external voltage required during operation is as low as 1 V and the HTR is only 10 s, which has a great potential for practical application.

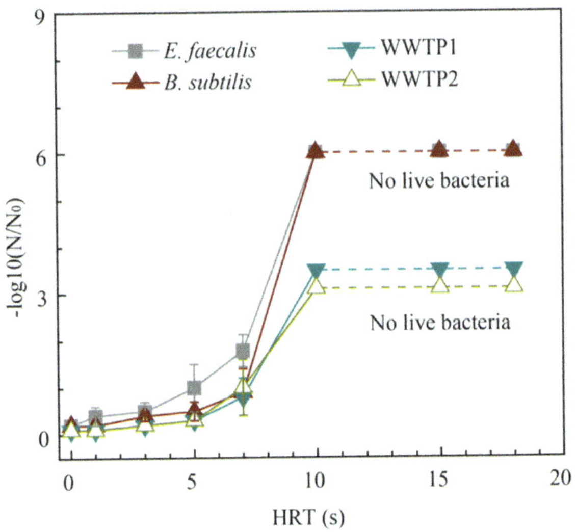

Fig. 2.5 Disinfection efficiency of CuO nanowire-assisted electroporation to *E. faecalis*, *B. subtilis*, and secondary effluent from municipal wastewater treatment plants. Dashed lines indicate that all microbes are inactivated, and no live microbes can be detected

2.4.3 *Current Fluctuations and Free Chlorine Generation During the Disinfection Process*

To investigate the disinfection by-product formation during nanowire-assisted electroporation, the current intensity in the circuit and the free chlorine generation during the treatment of actual water samples were measured. As shown in Fig. 2.6, the current intensity in the circuit was extremely low (1.2 mA) at a voltage of 1 V. At this time, no free chlorine was detected in the effluent after nanowire-assisted electroporation (detected limitation of 0.01 mg/L). This indicates that the nanowire-assisted electroporation with an external voltage of 1 V has a very low potential to generate chlorine-containing disinfection by-products during the treatment process. As the voltage increased, the current in the circuit gradually increased, and finally reached 65 mA at 5 V. The current intensity at this point was nearly 60 times higher than at 1 V. The free chlorine concentration in the effluent also increased with voltage. At 5 V, a free chlorine concentration of 0.36 mg/L was detected, and a large number of bubbles appeared in the effluent, indicating that the electrochemical reaction was significantly enhanced with the increasing voltage.

Benefiting from the designed porous electrodes, the nanowire-assisted electroporation disinfection technology can achieve chlorine-free and efficient microbial inactivation at an external voltage of 1 V. The extremely low external voltage not only reduces the electrochemical reaction during the treatment process but also prevents the generation of free chlorine and ensures that no disinfection by-products are generated in the effluent, which finally ensures the ecological safety of the treated water. In addition, due to the low external voltage, no electrolysis of the solution occurs during the treatment process, which ensures the energy efficiency of nanowire-assisted electroporation. The energy consumption required to treat the actual water sample is

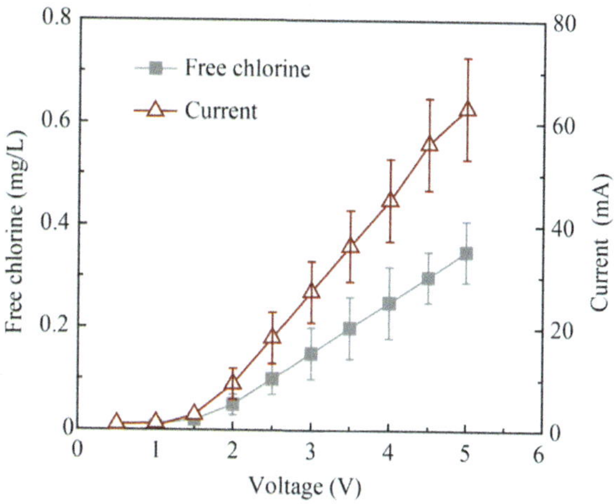

Fig. 2.6 Current fluctuations and free chlorine generation during the disinfection process

15 J/L under 1 V, 10 s operating conditions, which is the lowest energy consumption available for studies on disinfection technology and much lower than that required by conventional UV disinfection technology (about 1 × 10^3 J/L). The extremely low external voltage is ideal for application in small treatment units. Highly efficient microbial inactivation can be achieved driven by alkaline batteries, solar cells, and other power sources. By calculation, one commercial alkaline battery can treat more than 400 L of actual water.

2.5 Bacterial Inactivation Mechanisms of Nanowire-assisted Electroporation Disinfection

During nanowire-assisted electroporation disinfection, there was no significant bacterial interception during the disinfection process. In addition, the nanowire-bacteria contact and copper ion concentration in water did not contribute significantly to bacterial disinfection. Therefore, only electroporation of bacteria due to strong electric fields occurred during nanowire electroporation disinfection. To further validate the mechanism of nanowire-assisted electroporation disinfection, this study used SEM to analyze the morphology of bacteria before and after disinfection and also used fluorescence staining technique to analyze the structure of bacteria before and after disinfection.

2.5.1 Cell Morphology Analysis

The morphology of *E. coli* before and after nanowire-assisted electroporation disinfection was shown in Fig. 2.7. Before treatment, *E. coli* had a complete and smooth cell membrane structure without wrinkling. When *E. coli* were passed through the nanowire-assisted electroporation device at an HRT of 7 s without electricity, the cell membrane structure was still smooth and complete. This confirmed that the disinfection device had no significant inactivation effect on the bacteria in the absence of electricity.

When an external voltage of 1 V was applied, *E. coli* passed through the nanowire-assisted electroporation disinfection device at an HRT of 7 s. The membrane structure was significantly broken, as evidenced by the perforation of the cell membrane surface and the crumpling of the cell structure. Under this operating condition, *E. coli* was completely inactivated, confirming that the damage to the cell membrane was the mechanism of nanowire-assisted electroporation disinfection. Meanwhile, when the *E. coli* suspensions were treated using nanowire-assisted electroporation with other lethal dosages (e.g., 2 V, 5 s; 3 V, 2 s and 5 V, 1 s), the *E. coli* cell membrane surfaces were significantly broken. All the above *E. coli* morphological analyses indicated that the cell membrane of *E. coli* was damaged during nanowire-assisted electroporation disinfection, and the cell membrane breakage was the mechanism of cell death.

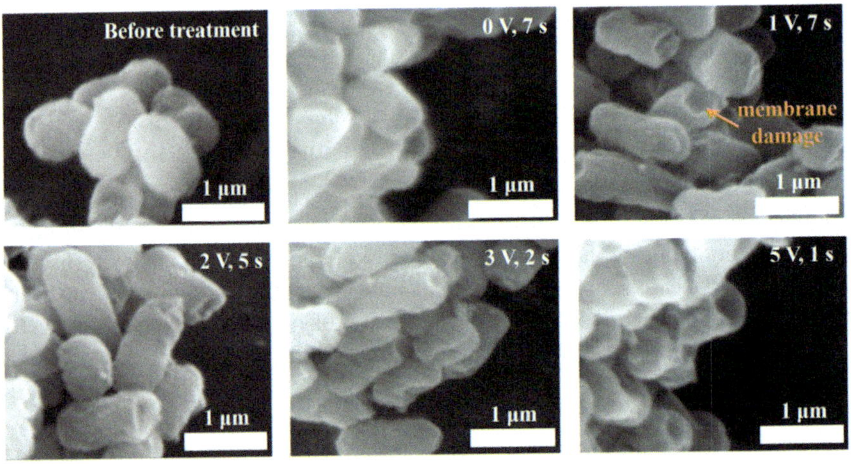

Fig. 2.7 Analysis of cell morphology after nanowire-assisted electroporation disinfection

2.5.2 Bacterial Staining Analysis

To further demonstrate electroporation as a disinfection mechanism, the study applied the cell staining method to analyze the cytoarchitecture of *E. coli* before and after treatment. Syto9 and PI are two types of nucleic acid dyes, of which Syto9 can cross the cell membrane, bind to nucleic acids in the cell, and emit green fluorescence under fluorescence excitation, as such Syto9 can label all cells. The PI cannot cross the active cell membrane with complete structure but can only enter the cell from the broken cell membrane Thus, the PI can label the cells with broken cell membrane and inactive cells [149].

The results of *E. coli* staining before and after nanowire-assisted electroporation disinfection treatment using Live/Dead staining method were shown in Fig. 2.8. For the untreated *E. coli* (before treatment), only a small number (<5%) of bacteria were labeled by PI, indicating that most of the bacteria had intact cell membrane structures and were cellularly active before treatment. After 1 V, 7 s electroporation disinfection, 90% of *E. coli* were PI-labeled, demonstrating that the cell membrane of bacteria was disrupted after nanowire-assisted electroporation disinfection treatment, leading to cell death. The above results demonstrated that the cell membrane of *E. coli* was disrupted during the nanowire electroporation disinfection process, and eventually led to cell death.

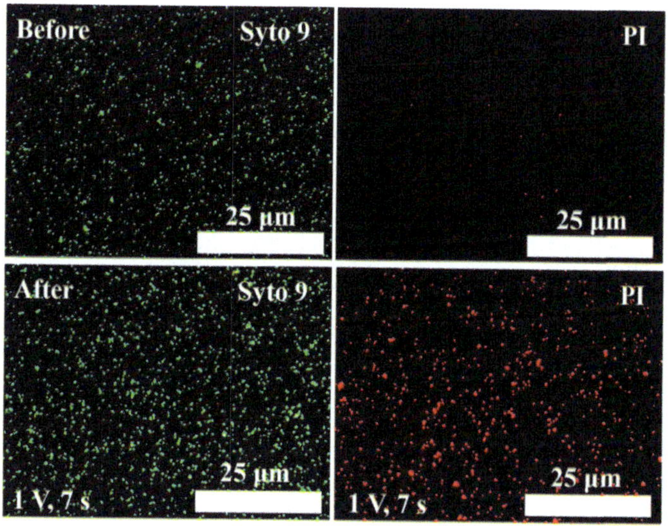

Fig. 2.8 Analysis of cell structure after nanowire-assisted electroporation disinfection

2.5.3 Bacterial Population Fluctuations During the Storage Process After Disinfection

In the reuse process, reclaimed wastewater needs to be transported and distributed to customers through the pipe network or stored in natural water bodies such as rivers and lakes for further ecological purification. To ensure the safety of wastewater reclamation, it is necessary to control the concentration of bacteria during the storage and distribution. For commonly applied chlorine disinfection technology, a certain concentration of residual chlorine is required in the actual treatment process to control the regrowth of microbes and ensure ecological safety. However, residual chlorine in reclaimed water can react with organic matter to produce carcinogenic chlorine-based disinfection by-products. In the case of UV disinfection technology, bacteria will reactivate and regrow in the subsequent storage and distribution process after UV disinfection. UV disinfection does not have a continuous bacterial disinfection effect. Therefore, reclaimed water disinfection technology needs to have the property of inhibiting microbes from reactivation and/or regrowth during storage and distribution process afterward. However, existing studies have not investigated the change of bacterial population during storage after nanowire-assisted electroporation disinfection. Therefore, in this study, we systematically analyzed the change of bacterial population during storage after nanowire-assisted electroporation disinfection and investigated the effect of treatment doseage and storage substrate on the change of bacterial population.

2.5.4 Bacterial Population Fluctuations During the Storage Process

(1) Investigation of bacterial population changes during storage after nanowire-assisted electroporation disinfection using lethal dosage

In this study, after treated with a lethal dosage (1 V, 7 s), *E. coli* were stored in different substrates (DI water, saline, and filtered secondary effluent) to study the reactivation and regrowth phenomenon of dead bacteria during storage. The inactivated *E. coli* were stored under light and their numbers changed as shown in Fig. 2.9. In different storage substrates, no viable bacteria were detected (detection limit < 10 CFU/mL) during 24 h of storage for the bacteria treated with lethal dosage. This demonstrates that nanowire-assisted electroporation disinfection ensures that dead bacteria do not reactivate and regrow during storage and ensures the ecological safety of reclaimed water reuse.

(2) Investigation of bacterial population changes during storage after nanowire-assisted electroporation disinfection using non-lethal dosage

The study further analyzed the inhibitory effect of nanowire-assisted electroporation disinfection on the inactivated bacteria during the storage process. Bacteria treated

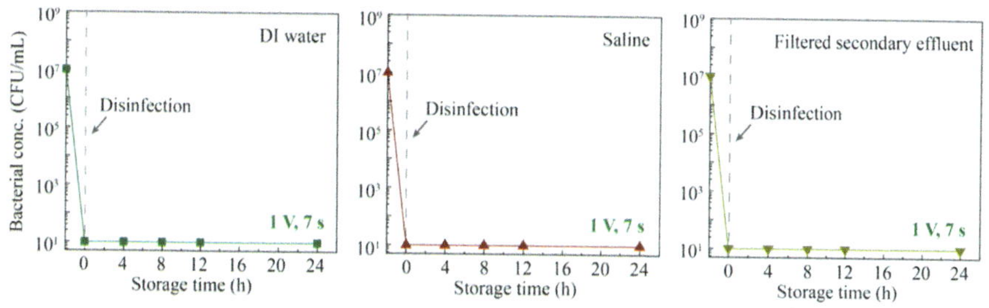

Fig. 2.9 Investigation of bacterial population changes during storage after nanowire-assisted electroporation disinfection using lethal dosage

with non-lethal dosage (1 V, 3 s and 1 V, 5 s) were stored in different substrates and the change in bacterial population during the storage was analyzed. As shown in Fig. 2.10, the concentration of active bacteria in the effluent after the 1 V, 3 s treatment was the same as that before the treatment. This indicated that the lower treatment dosage was not able to inactivate *E. coli* immediately after treatment. However, during the subsequent storage process, the number of live *E. coli* decreased significantly in different storage substrates. The most significant decrease of active bacteria was observed in the DI water, where no viable bacteria were detected in the DI water after 24 h of storage. The rate of decline in the number of live bacteria in the secondary effluent was lower than that in DI water, with a concentration of about 10^2 CFU/mL after 24 h. The slowest decline of live bacteria was observed in saline, with a concentration of 10^4 CFU/mL still detected after 24 h of storage.

The difference in the number of live bacteria stored in different substrates after treatment with the same disinfection dosage was likely due to different the osmotic pressure between the bacteria and the substrate: the higher the osmotic pressure, the faster the rate of bacterial death. After treatment, the surface of the bacteria is

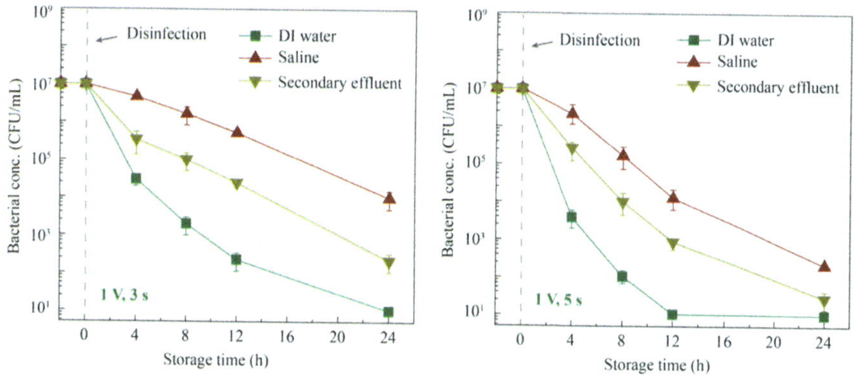

Fig. 2.10 Investigation of bacterial population changes during storage after nanowire-assisted electroporation disinfection using non-lethal dosage

damaged and the osmotic pressure between the bacteria and the substrate may accelerate the efflux of cellular contents and intensify the destruction of cell membranes, eventually leading to cell death. Therefore, for cells stored in DI water, the number of live cells decreases fastest due to the maximum osmotic pressure between the cytoplasm and the solution. While when stored in saline, the number of live cells decreases the slowest due to the lowest osmotic pressure.

We further investigated the concentration change of live bacteria stored in different substrates after nanowire-assisted electroporation disinfection with 1 V, 5 s (another non-lethal dosage). The change of live bacteria during the storage process was similar to those after the 1 V, 3 s treatment. The bacteria were still active after treatment., whereas, during storage, the number of live bacteria gradually decreased with incubation time. For the same substrate, the rate of decrease in the number of live bacteria after 1 V, 5 s treatment was faster than that after 1 V, 3 s treatment. After 1 V, 5 s treatment and stored in DI water for 12 h, no live bacteria can be detected. For saline, the concentration of live bacteria was 10^2 CFU/mL after 1 V, 5 s treatment with 24 h storage. Compared to the HRT of 3 s, the 5 s treatment dosage further disrupted the cell membrane structure and accelerated the rate of reduction of live bacteria during the storage process. The rate of decrease in the number of live bacteria during storage after 1 V, 5 s treatment was also influenced by the osmotic pressure of the cytoplasm and the storage substrate. The rate of decrease of live bacteria when stored in DI water was greater than that in filtered secondary effluent, while the rate of decrease of live bacteria when stored in saline was the lowest. In addition, without a power supply, *E. coli* passing through the device were also collected and stored in different substrates, and the changes in the number of live bacteria were measured during the storage process. As shown in Fig. 2.11, the number of live bacteria did not change significantly when the bacteria were stored in different substrates for 24 h. This indicated that the bacteria passed through the reactor without power supply showed no significant changes in bacterial numbers during the subsequent storage process and the cell activity did not change significantly over a certain time when stored in different substrates.

In summary, nanowire-assisted electroporation disinfection not only ensures that the inactivated bacteria do not reactivate during the storage process but also inhibits the regrowth of the treated bacteria and promotes their death during storage, which can effectively control the ecological risk during the reuse of reclaimed water.

2.5.5 *Structural Analysis of Bacterial Morphology During Storage After Low-Dosage Nanowire-assisted Electroporation Disinfection*

During storage after low-dosage disinfection treatment (1 V, 3 s), the number of live bacteria showed a significant trend of decline. To study the mechanism of bacterial death during storage, cell staining experiments were performed to analyze the

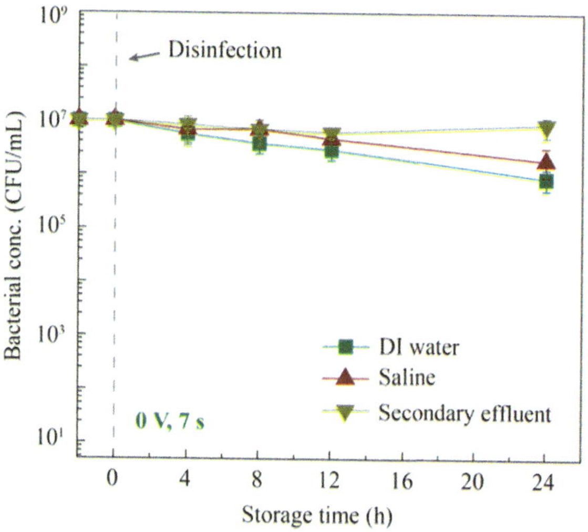

Fig. 2.11 Investigation of bacterial population changes during storage without treatment

changes in cell structure during storage. As shown in Figs. 2.12 and 2.13, the bacterial cells before treatment were unstained by PI dye (~5%) without significant membrane structure breakage, and some cells after the treatment immediately (0 h) (~20%) were labeled by PI, indicating that a certain number of cells already showed significant cell membrane breakage. The percentage of cells stained with PI dye increased with the storage time. After 4 h of storage, 95% of the cells were stained with PI, and when the storage time was increased to 24 h, 98% of the cells were stained with PI.

The above results indicate that the cell membrane structure has been disrupted during storage after low-dosage nanowire-assisted electroporation disinfection. However, the initial damage to the cell membrane was trace and tiny due to the low dosage of electroporation disinfection treatment, so only some cells were labeled by PI dye immediately after disinfection. During the storage process afterward, the damage on the cell membrane became larger. After 4 h, the cell membrane of almost

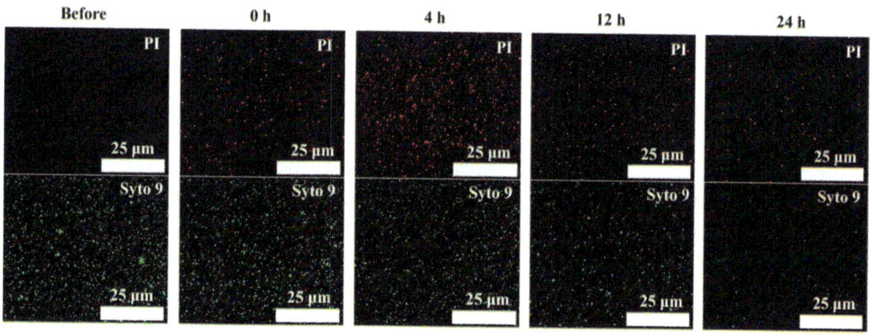

Fig. 2.12 Analysis of bacterial staining during storage after low-dosage nanowire-assisted electroporation disinfection

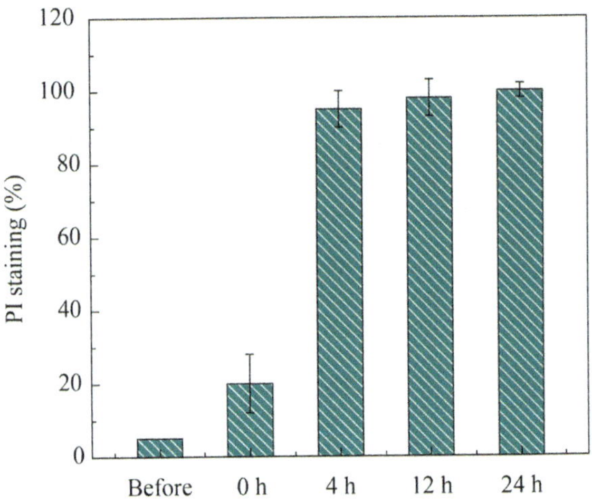

Fig. 2.13 Proportion of bacterial staining during storage after low-dosage nanowire-assisted electroporation disinfection

all cells showed significant breakage and was labeled by PI dye. In addition, the total number of cells labeled by the Syto9 dye also decreased continuously after a longer storage time (24 h). The possible reason for this is that with the increase of storage time, some cells showed more obvious breakage and were unable to maintain their cell morphology, resulting in the inability to be labeled by Syto9 dye. The above cell staining experiments indicated that during storage after low-dosage nanowire-assisted electroporation disinfection, the breakage in the cell membrane due to electroporation was increasing. The process of increasing breakage led to continuous cell death during storage.

In addition, the study further used SEM to investigate the morphological changes of the cells during storage after the low-dosage treatment and to analyze the cell inactivation mechanism during storage. As shown in Fig. 2.14, in contrast to the smooth and intact cell membrane before electroporation, some wrinkling occurred on the surface of the bacteria after low-dosage (1 V, 3 s) treatment. Although the surface was not significantly perforated and broken, the more wrinkled cell surface proved that the bacteria were damaged. In contrast, after 4 h storage, significant cell membrane breakage appeared on the bacterial surface, visually verifying the increased cell membrane damage during storage. As the storage time increased, the damage on the cell membrane also increased. After 12 h storage, some of the cells were not able to maintain the normal cell structure, and extensive cell membrane disruption and cell collapse occurred. After 24 h storage treatment, the cells completely lost their cell structure. The above results also confirmed that the cell membrane breakage continued to increase during storage, leading to continuous cell inactivation.

Fig. 2.14 Analysis of bacterial morphology during storage after low-dosage nanowire-assisted electroporation disinfection

2.6 Summary of the Tendency of Bacterial Changes During Storage After Disinfection

The tendency of bacterial changes during storage after nanowire-assisted electroporation disinfection is summarized as follows (Fig. 2.15).

(1) Lethal dosage

Cells treated by nanowire-assisted electroporation disinfection using a lethal dosage (1 V, 7 s) received severe damage to the cell membrane due to significant electroporation (Figs. 2.7 and 2.8). None of these cells reactivated during the subsequent storage process (Fig. 2.9).

(2) Low-dosage

After the cells were treated by nanowire-assisted electroporation disinfection using a low-dosage (1 V, 3 s and 1 V, 5 s), the damage to the cell membrane surface was relatively mild due to the lowered treatment dosage. No significant cell membrane breakage was observed by SEM right after treatment, and only some cells were labeled with PI dye using cell staining. Due to the relatively mild cell damage, the cells were still able to count using plate culture and exhibited cellular activity. During subsequent storage, the cell membrane breakage gradually increased, and the degree of cell membrane breakage was also correlated with the osmotic pressure between the cytoplasm and storage substrates. After a period of storage, the cells were unable to exhibit normal cellular activity due to severe breakage on the membrane and were not able to count using the plate culture method.

The above results demonstrate that nanowire-assisted electroporation disinfection technology can effectively inhibit the growth of bacteria that are still active after treatment and can promote their inactivation during the storage afterward, effectively controlling the ecological risk after disinfection.

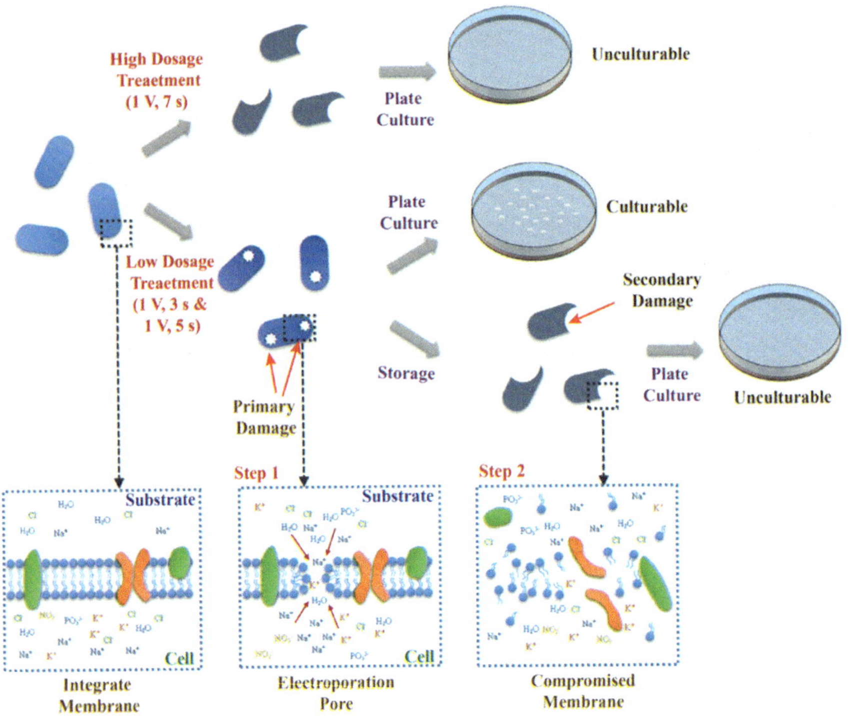

Fig. 2.15 Schematic showing the change live bacteria during storage after nanowire-assisted electroporation disinfection

2.7 Summary

(1) Using the thermal oxidation method to prepare CuO nanowire-modified copper foam electrodes can allow bacterial suspensions or actual water samples to pass through the interior of the porous electrode during operation. This flow-through operation method effectively increases the opportunities of bacteria approaching the nanowire-enhanced strong electric field region near the electrode surface, which enhances disinfection efficiency.

(2) The operating conditions required for complete microbial inactivation using the flow-through electroporation disinfection with CuO nanowire-modified copper foam porous electrodes were an external voltage of 1 V and a contact time (HRT) of less than 10 s. An operating voltage of 1 V effectively reduced the potential for chlorine-based disinfection by-products and effectively reduced operating energy consumption. HRT of 10 s ensures the efficient operation of nanowire-assisted electroporation disinfection.

(3) Cell morphology and cell staining analysis indicated that the disinfection mechanism was electroporation, and no chemicals were needed during the treatment process.

(4) Analysis of the fluctuations of live bacteria during storage after electroporation disinfection indicated that the nanowire-assisted electroporation disinfection technology not only ensured no reactivated of dead cells but also inhibits the regrowth of inactivated bacteria and promotes their death during storage. It is proved that electroporation disinfection can effectively reduce the ecological risk during storage.

Chapter 3
Effect of the Nanowire Morphology and Electrode Structure on Microbial Inactivation

3.1 Research Background

Nanowire morphology and electrode structure directly affect the inactivation efficiency of bacteria during nanowire-assisted electroporation disinfection. It has been shown that the aspect ratio (length:diameter) of nanowires is directly related to the electric field enhancement effect [150, 151]. The higher the aspect ratio, the more elongated the nanowires appear, and the more pronounced the local electric field enhancement effect is. Therefore, theoretically, a high aspect ratio has a beneficial effect on bacterial inactivation. However, the preparation of nanowires with high aspect ratios usually requires more demanding operating conditions, which not only increases the energy consumption, but also the physical structure of nanowires with high aspect ratios is usually fragile and easily breaks under the impact of water flow. The easily detached nanowires cause the loss of heavy metals in the electrode, which increases the ecological risk of disinfected water and affects the service life of the electrode to ensure the long-term operating performance. Up to date, there are no studies on the effect of nanowire morphology on nanowire-assisted electroporation disinfection. In addition, there is no research on the effect of electrode structure on the efficiency of nanowire-assisted electroporation disinfection. Theoretically, reducing the electrode pore size or increasing the electrode thickness can increase the opportunities of bacteria approaching the electrode surface, thus increasing the efficiency of nanowire-assisted electroporation disinfection. However, both smaller pore sizes and thicker electrodes are more likely to cause clogging of particulate matter in the water, which affects the disinfection effect. In addition, there may be different contributions of positive and negative electrodes to disinfection during nanowire electroporation, so the positive and negative electrode designs need to be optimized, but the related studies are still limited. In summary, studying the effect of nanowire morphology and electrode structure on microbial inactivation is important for the design and operation of nanowire-assisted electroporation disinfection reactors.

Z. Huo, *Nanowire-assisted Flow-through Electrode Enabling Electroporation Disinfection of Reclaimed Water*, Springer Theses,
https://doi.org/10.1007/978-981-99-4502-3_3

The objectives of this chapter are to clarify the means of in-situ growth of CuO nanowires with different morphologies on copper foam surfaces. Investigate the impacts of nanowire morphology and electrode structure on the disinfection effect of nanowire-assisted electroporation disinfection. Analyze the respective contributions of positive and negative electrodes in the actual treatment process. Finally, optimize the design of nanowire-assisted electroporation electrodes according to the above impact factors to provide a theoretical basis for reactor design.

3.2 Experimental Materials and Methods

3.2.1 Experimental Reagents

The reagents are the same as in Sect. 2.2.1.

3.2.2 Preparation of Porous Electrodes Modified with Nanowires of Different Morphologies

(1) Preparation of nanowires with different morphologies

Copper foam with a diameter of 10 mm, thickness of 5 mm, and pore size of 500 μm was selected as the porous electrode in the nanowire-assisted electroporation disinfection device used in Sect. 2.2.1. CuO nanowires were prepared on the surface of the copper foam electrode using the thermal oxidation method used in Sect. 2.2.1. As shown in Table 3.1, the operating conditions (heating rate, heating temperature, and heating time) were adjusted during the preparation process to prepare CuO nanowires with different morphologies.

(2) Preparation of CuO nanowire-modified copper foam electrodes with different electrode structures

Copper foam with different electrode structures was purchased from Jiayisheng Electronics Co., Ltd. in Kunshan, China, and the manufacturer processed the copper foam into a cylindrical electrode with a diameter of 10 mm using laser cutting. As shown in Table 3.2, CuO nanowire-modified copper foam electrodes with different electrode structures (pore size and electrode thickness) were prepared using the thermal oxidation method used in Sect. 2.2.1. The controlled electrode preparation conditions were heated to 400 °C at a heating rate of 3.3 °C/min and kept heated at 400 °C for 2 h.

Table 3.1 Operating conditions for fabricating the CuO nanowires by thermal oxidation

Heating temp. (°C)	Heating rate (°C/min)	Heating time (min)	Heating temp. (°C)	Heating rate (°C/min)	Heating time (min)
350	3.3	120	450	3.3	180
400	3.3	120	500	3.3	180
450	3.3	120	550	3.3	180
500	3.3	120	350	1.1	120
550	3.3	120	400	1.1	120
600	3.3	120	450	1.1	120
350	3.3	90	500	1.1	120
400	3.3	90	550	1.1	120
450	3.3	90	350	2.2	120
500	3.3	90	400	2.2	120
550	3.3	90	450	2.2	120
350	3.3	150	500	2.2	120
400	3.3	150	550	2.2	120
450	3.3	150	350	4.4	120
500	3.3	150	400	4.4	120
550	3.3	150	450	4.4	120
350	3.3	180	500	4.4	120
400	3.3	180	550	4.4	120

Table 3.2 CuO nanowire-modified copper foam electrodes with different electrode structures

Pore size (μm)	Thickness (mm)	Pore size (μm)	Thickness (mm)	Pore size (μm)	Thickness (mm)
100	5	100	10	100	2
150	5	150	10	150	2
200	5	200	10	200	2
300	5	300	10	300	2
500	5	500	10	100	1
700	5	700	10	150	1
1000	5	1000	10	200	1

3.2.3 *Construction of Nanowire-assisted Electroporation Disinfection Devices with Different Electrode Structures*

The disinfection device used for the study was based on the device used in Sect. 2.2.2 with appropriate modifications. The prepared electrodes with different morphology of CuO nanowires were placed in the corresponding Plexiglas electrode holders of different sizes. The electrodes were separated by a plastic mesh (200 μm pore size) to prevent short circuits. The two electrodes were connected to the DC power supply (Beijing Dahua, DH1715A) through wires to serve as the positive and negative electrodes, respectively.

3.2.4 Characterization of CuO Nanowire-Modified Copper Foam Electrode

(1) Morphological characterization of CuO nanowires

The CuO nanowires-modified copper foam electrodes were characterized using the SEM characterization technique used in Sect. 2.2.3.

(2) Statistical analysis of CuO nanowire morphology

The image analysis software Image-Pro Plus 6.0 was used to analyze the morphology of the CuO nanowires under different operating conditions. The SEM photographs were loaded onto Image-Pro Plus 6.0, and the lengths and diameters of the nanowires were measured. The mean and standard deviation of the lengths and diameters of the prepared nanowires under each operating condition were calculated using SPSS 22.0 software.

3.2.5 Microbes and Water Samples Used in Experiments

(1) Microbes used in experiments

E. coli (CGMCC 1.3373), the culture method is the same as in Sect. 2.2.4.

(2) Water samples used in experiments

E. coli suspension, the preparation method is the same as in Sect. 2.2.4.

3.2.6 Nanowire-assisted Electroporation for Microbial Disinfection

(1) HRT calculation

The calculation method is the same as in Sect. 2.2.5. To consider the effect of electrode thickness on nanowire-assisted electroporation disinfection and to compare the bacterial inactivation effect under the same HRT, different flow rates under the same HRT were calculated separately for different sizes of electrodes.

(2) Nanowire electroporation disinfection experiment

The experimental method is the same as in Sect. 2.2.5. The applied external voltage is 1 V and HRT is 7 s.

(3) Bacteria counting and nanowire-assisted electroporation disinfection performance evaluation method

The experimental method is the same as in Sect. 2.2.5.

3.2.7 Investigation of the Disinfection Contribution of Positive and Negative Electrode and Optimization of the Reactor Design

(1) Investigation of the disinfection contribution of positive and negative electrode

A copper foam electrode with a pore size of 500 μm and a thickness of 5 mm was selected and heated to 400 °C at a heating rate of 2.2 °C/min and kept heated at 400 °C for 2 h to prepare an effective disinfection electrode. The copper foam electrode with the same structure was selected and heated to 600 °C at a heating rate of 20 °C/min and kept heated at 600 °C for 2 h. The electrode surface without nanowire structure was obtained using this method, which did not contribute to electroporation disinfection. The fabricated ineffective electrodes were set as a control electrode [77].

First, the positive electrode was set as the effective electrode and the negative electrode as the control electrode. The disinfection performance was tested by diluting *E. coli* into DI water or 10 mM NaCl solution with an external voltage of 1 V and an HRT of 7 s.

Then, the negative electrode was set as the effective electrode and the positive electrode as the control electrode. The disinfection performance was tested by diluting *E. coli* into DI water or 10 mM NaCl solution with an external voltage of 1 V and an HRT of 7 s.

The disinfection effect of positive and negative electrodes was compared and the contribution of positive and negative electrodes in the disinfection process was analyzed.

To clarify the reasons for the different contributions of positive and negative electrodes to disinfection in actual water bodies, bacteria were diluted into NaCl solution, and the pH of the solution was adjusted using HCl and NaOH, and the final solution ionic strength was maintained at 10 mM. The microbial disinfection efficiency by positive and negative electrodes was measured at different pH values. The external voltage was set at 1 V and the HRT was 7 s. The surface charge distribution of bacteria in the solution was also measured at different pH values.

The bacterial surface charge distribution was determined as follows: *E. coli* cultured to logarithmic growth period (12 h) were diluted into NaCl solutions at

Table 3.3 Effect of different combinations of electrode pore sizes on nanowire-assisted electroporation disinfection

	Positive electrode	Negative electrode
Pore size (μm)	200	500
Pore size (μm)	500	200
Pore size (μm)	500	500
Pore size (μm)	200	200

different pHs while maintaining the ionic strength at 10 mM and keeping the bacterial concentration at 10^5 CFU/mL. The bacterial zeta potential was measured using a high-sensitivity nanoparticle size analyzer (Beckman, DelsaNano C). Each group of samples was repeated five times, the mean value was taken, and the standard deviation was calculated.

(2) Reactor optimization design

As shown in Table 3.3, different pore sizes of electrodes were applied, and *E. coli* was diluted into 10 mM NaCl solution to test the disinfection effect under different combinations of electrode pore sizes for reactor optimization design.

3.3 Investigation on the Effect of CuO Nanowire Morphology on Bacterial Disinfection

3.3.1 Factors Impacting the Morphology of CuO Nanowires

The preparation of CuO nanowires on the surface of copper foam by thermal oxidation consists of three main processes. As shown in Fig. 3.1: (1) heating stage: heating the electrode at a certain rate so that the electrode reaches the specified temperature within a preset time; (2) constant temperature heating stage: after the heating process, the electrode is kept at the specified temperature for a certain time; (3) cooling stage: after the specified time of constant temperature heating, the heating is stopped, and the electrode will cool naturally to room temperature. Therefore, during the preparation process, the heating temperature, the heating time, and the heating rate will all impact the morphology of the CuO nanowires.

(1) Heating temperature

The effect of different heating temperatures on the morphology of the nanowires was investigated by fixing the heating rate and the heating time at 3.3 °C/min and 120 min, respectively. As shown in Fig. 3.2, the heating temperature significantly affects the morphology of the CuO nanowires. The nanowire length tends to increase and then decrease with the increase of heating temperature, while the nanowire diameter increases with the increase of heating temperature.

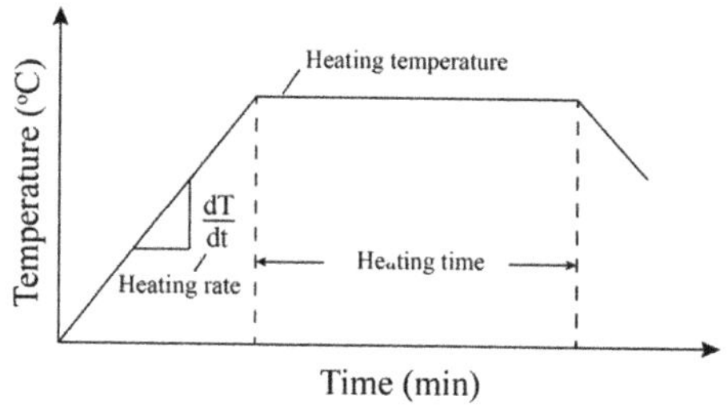

Fig. 3.1 Schematic showing each stage of CuO nanowire preparation by thermal oxidation

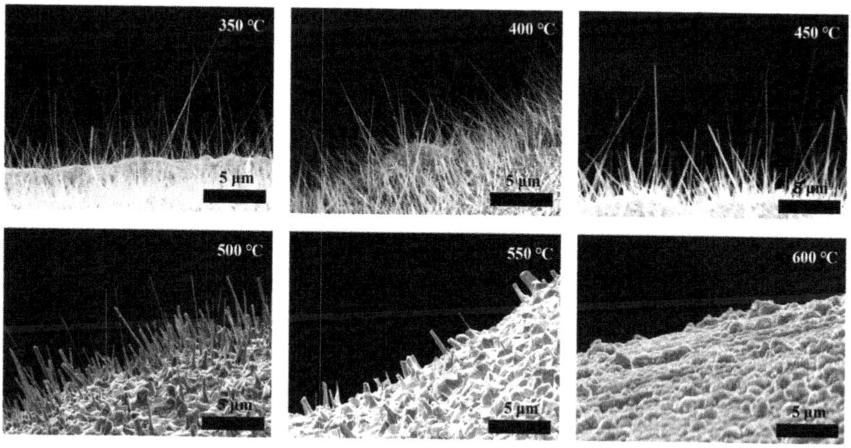

Fig. 3.2 Investigation of different heating temperatures on CuO nanowire morphology

As shown in Fig. 3.3, after heating at 350 °C, CuO nanowires grew on the surface of the copper foam electrode with a length of 6.2 μm and a diameter of 20 nm. When heated at 400 °C, the length of CuO nanowires increased to 15.1 μm, while the diameter of nanowires increased to 40 nm. Increasing the temperature, the length of nanowires decreased, while the diameter increased significantly. After heating at 450, 500, and 550 °C, the average length of nanowires decreased to 9.2, 6.5 and 1.1 μm, respectively, while the average diameter increased to 72, 150 and, 300 nm.

(2) Constant heating time

This study also investigated the effect of constant heating time on nanowire morphology. Considering no CuO nanowire was generated on the surface of the copper foam electrode after heating at 600 °C, the specific experimental conditions were as follows: the heating temperature was set at 350, 400, 450, 500, and 550 °C, and the heating time (90, 150, and 180 min) was applied at each heating temperature.

The nanowire morphology at different heating times was shown in Fig. 3.4. The length of the nanowires increased and then decreased with the increase of heating time, while the diameter of the nanowires increased with the increase of heating time.

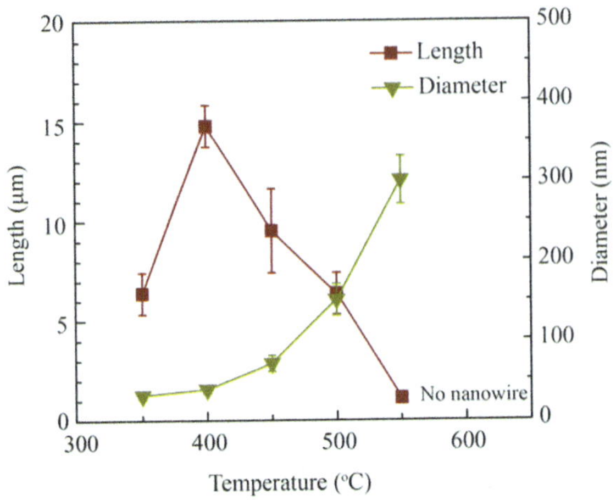

Fig. 3.3 Analysis of the impact of heating temperature on the morphology of CuO nanowires

When the heating temperature was 550 °C, the nanowire length decreased significantly due to the high temperature, and the nanowire structure was not significant at all heating times.

Figure 3.5 further analyzes the effect of constant heating time on the nanowire morphology. The nanowire length reached the longest after 120 min of heating treatment for all experimental groups with different heating temperatures. Except for the optimized heating time, the short or long heating time is not conducive to the growth of ideal-shaped CuO nanowires.

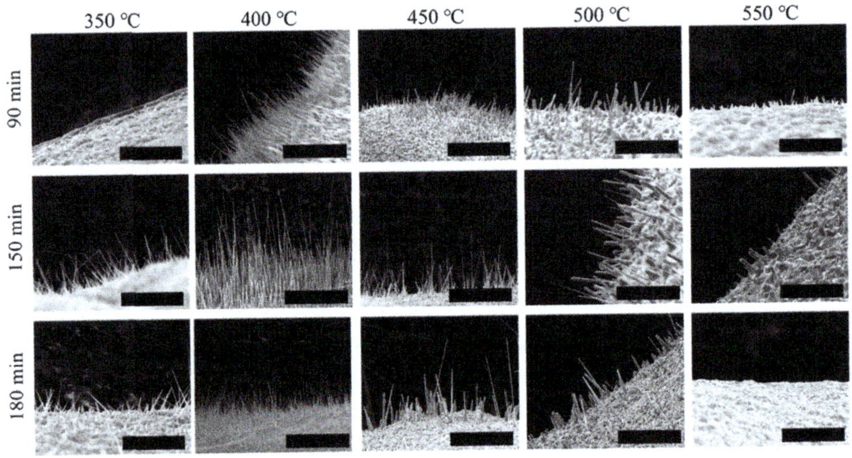

Fig. 3.4 Effect of heating time on CuO nanowire morphology. Scale bars are 5 μm

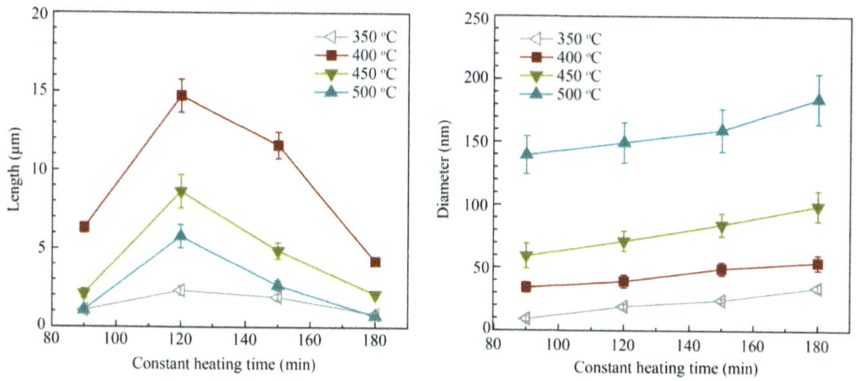

Fig. 3.5 Analysis of the impact of heating time on the morphology of CuO nanowires

(3) Heating rate

Heating rate is another important impact factor on CuO nanowire morphology during thermal oxidation. According to the above results, the CuO nanowires had the longest length at the heating time of 120 min. Therefore, the specific experimental conditions were as follows: the heating temperature was set at 350, 400, 450, 500, and 550 °C, and the heating time was kept constant at the corresponding heating temperature (120 min). The heating rate was adjusted at different rates (1.1, 2.2, 3.3, and 4.4 °C/min) to investigate its effect on the morphology of CuO nanowires. As shown in Fig. 3.6, when the heating rate was 1.1 °C/min, the CuO nanowires had an ideal shape at each heating temperature, and when the heating rate was increased to 2.2 °C/min, the length and diameter of the CuO nanowires increased. When the heating rate was further increased to 3.3 °C/min, the length and diameter of CuO nanowires decreased, and the nanowire structure was not obvious. At the final heating rate of 4.4 °C/min, no obvious nanowire structure was generated at each heating temperature.

The effect of the thermal oxidation process including heating temperature, constant heating time, and heating rate on CuO nanowires can be explained by its growth mechanism. The fabrication of CuO nanowires by thermal oxidation is divided into two main processes: (1) The copper substrate first reacts with oxygen in the air to form a thin film of cuprous oxide (Cu_2O) on the surface. Due to the reaction between oxygen and copper, a region with low oxygen concentration is formed near the surface of the copper substrate; (2) during the gradual heat process, Cu_2O further reacts with oxygen in the air to produce CuO. In the process of reaction between Cu_2O and oxygen, the oxygen in the air needs to diffusion to the copper surface. The diffusion direction of oxygen is perpendicular to the substrate surface, and the

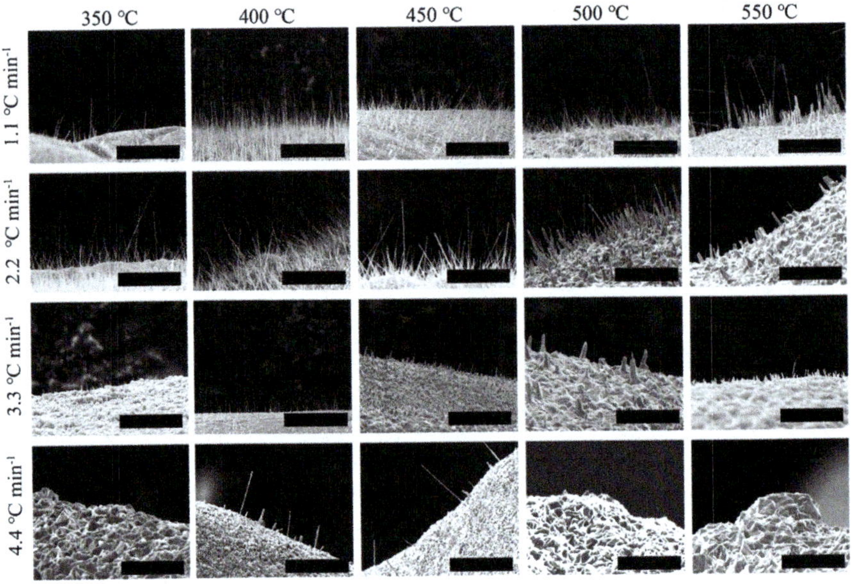

Fig. 3.6 Effect of heating rate on CuO nanowire morphology. Scale bars are 5 μm

reaction of Cu_2O with oxygen will occur in the opposite direction of oxygen diffusion, which produces CuO nanowires [103]. By adjusting the appropriate operating conditions, CuO nanowires with ideal morphology can be prepared.

During the process of CuO nanowire fabrication, the heating temperature directly affects the morphology and structure of the synthesized material [103]. When the heating temperature is lower than 350 °C, the oxygen diffusion rate is too slow to synthesize the nanowire structure. As increasing the heating temperature to 400–450 °C, the oxygen diffusion rate can be accelerated, and the length of CuO nanowires can be increased. However, when the heating temperature is too high and the oxygen diffusion rate increases significantly, the length of nanowires on the surface of the copper substrate will decrease and the diameter will increase further. The effect of constant heating time and heating rate on nanowire morphology is similar to that of heating temperature, i.e., it mainly affects the oxygen diffusion rate. With the increase of heating time and heating rate, the length of CuO nanowires showed a tendency to increase and then decrease, while the diameter increased. In addition, when the heating temperature was higher than 964 °C, the reaction between Cu_2O and copper no longer formed CuO. Therefore, under the operating conditions used in this study, the nanowire structures generated were all CuO nanowires with relatively similar electrical conductivity, strength, and corrosion resistance. Therefore, the properties of the CuO nanowires synthesized under different preparation conditions are comparable in the subsequent studies.

3.3.2 Study on the Impact of CuO Nanowire Morphology on Bacterial Disinfection

To investigate the impact of different CuO nanowire morphologies on bacterial disinfection, copper foam electrodes modified with nanowires with different morphologies were prepared and the disinfection efficiency of different electrodes was evaluated.

(1) Nanowire length

The effect of different nanowire lengths on the disinfection efficiency of nanowire-assisted electroporation disinfection was analyzed. The disinfection efficiency of nanowire-modified copper foam electrodes was measured with an external voltage of 1 V and an HRT of 7 s. The results are shown in Fig. 3.7. The disinfection efficiency of nanowire-assisted electroporation increases with the increase of nanowire length. When the nanowire length was below 3.8 μm, the nanowire-assisted electroporation disinfection efficiency was low (log removal rate <1.0 log). As the nanowire length above 3.8 μm, the electroporation disinfection efficiency increased significantly with the increase of the nanowire length and achieved complete microbial inactivation (log removal rate >6.0 log) when the nanowire length reached 6.4 μm. When the nanowire length was higher than 6.4 μm, the electroporation disinfection device can achieve the complete microbial disinfection.

(2) Nanowire diameter

The effect of nanowire diameter on the disinfection effect of nanowire-assisted electroporation was further analyzed. The results are shown in Fig. 3.8. The disinfection efficiency of nanowire-assisted electroporation decreases with the increase of

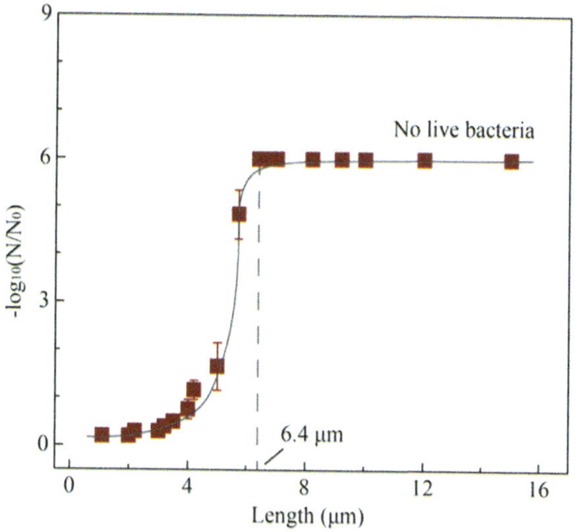

Fig. 3.7 Effect of nanowire length on electroporation disinfection performance

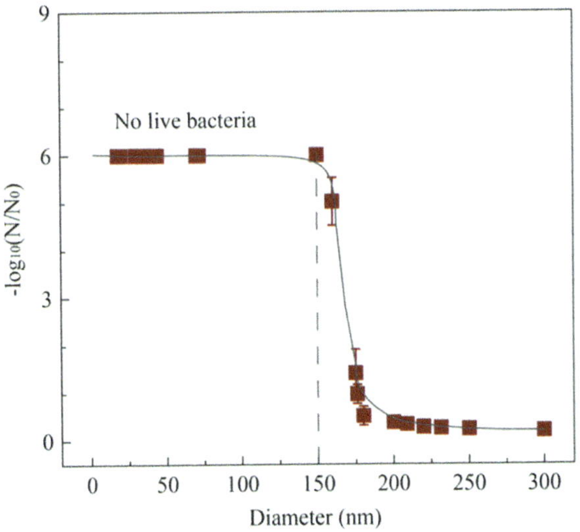

Fig. 3.8 Effect of nanowire diameter on electroporation disinfection performance

nanowire diameter. When the nanowire diameter was less than 150 nm, nanowire-assisted electroporation can achieve complete microbial disinfection (log removal rate >6.0 log). As the nanowire diameter greater than 150 nm, the nanowire-assisted electroporation disinfection efficiency decreases significantly with the increase of nanowire diameter. When the nanowire diameter was greater than 175 nm, the nanowire-assisted electroporation disinfection has almost no microbial inactivation effect (log removal rate <1.0 log).

(3) Nanowire aspect ratio

Since the thermal oxidation fabrication process has a similar effect on the length and diameter of the nanowire morphology and the nanowires with different sizes were prepared in a limited number, the analysis of the effect of nanowire length or diameter on the electroporation disinfection efficiency is thus limited. The effect of nanowire length or diameter on electroporation disinfection efficiency cannot be explained mechanistically and cannot reasonably guide the subsequent preparation and synthesis of other nanowire materials.

The aspect ratio of the nanowires was found to be positively correlated with the electric field enhancement by placing the nanowires perpendicularly on the electrode surface, which can be estimated by Eqs. (3.1) and (3.2).

$$\frac{E_{tip}}{E_0} = \alpha \cdot \frac{l_{rod}}{d_{rod}} \tag{3.1}$$

$$E_0 = \frac{U}{d} \tag{3.2}$$

where E_0 is the strength of the bulk electric field between two electrodes, α is a constant, l_{rod} and d_{rod} are the length and the diameter of the nanowire, U is the external voltage applied to electrodes, and d is the distance of the electrodes. Considering the electric field at the nanowire tip directly affects the electroporation disinfection efficiency, the nanowire aspect ratio was the main factor affecting the disinfection performance. Therefore, the effect of different nanowire aspect ratios on nanowire-assisted electroporation disinfection was analyzed to investigate the disinfection performance of nanowire-modified copper foam electrodes with different aspect ratios. The external voltage and HRT were fixed at 1 V and 7 s, respectively.

As shown in Fig. 3.9, the effect of nanowire aspect ratio on electroporation disinfection was significant. The nanowire-assisted electroporation disinfection efficiency increased with the nanowire aspect ratio. As the nanowire aspect ratio <20, electroporation had no inactivation of microbes (log removal rate <1.0 log). When the nanowire aspect ratio increased, the electroporation disinfection efficiency increased significantly and achieved complete inactivation of microbes at a nanowire aspect ratio of 40 (log removal rate >6.0 log). Further increased nanowire aspect ratio can remain high disinfection efficiency (complete inactivation with log removal rate >6.0 log). Thus, the nanowire aspect ratio is confirmed to be the main factor affecting the disinfection performance in the electroporation disinfection process. In the follow-up study, the nanowire diameter needs to be nanoscale and the aspect ratio is supposed to be greater than 40 to achieve efficient nanowire-assisted electroporation disinfection.

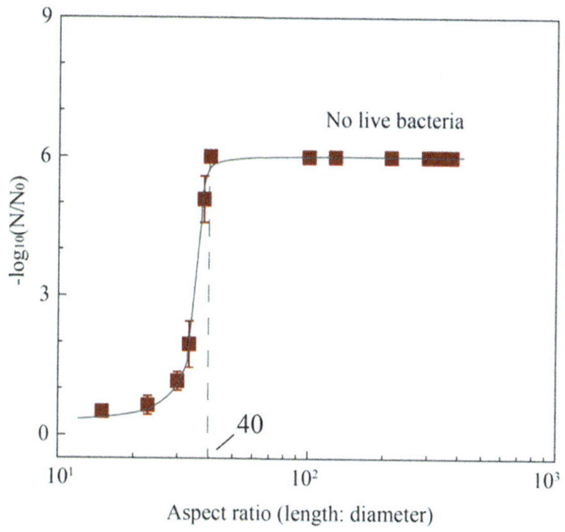

Fig. 3.9 Effect of nanowire aspect ratio on electroporation disinfection performance

3.4 Investigation on the Effect of Electrode Structure on Bacterial Disinfection

After clarifying the effect of nanowire morphology on electroporation disinfection, the study further analyzed the effect of porous electrode structures on bacterial inactivation efficiency. When copper foam was selected as the porous electrode for the reactor design, the electrode structures to be optimized were the electrode pore size and electrode thickness.

The prerequisite for achieving efficient electroporation disinfection is to enlarge the electric field enhancement area to reduce the time for microbes to approach the electrode surface. The smaller the electrode pore size, the faster the microbes approach the electrode surface. However, the pore size may bring about problems such as internal blockage of the porous electrode, affecting the reactor operation, as such its effect on electroporation disinfection efficiency need to be systematically studied. In addition, the greater the electrode thickness, the greater the probability that microbes in the water will come close to the electrode surface when passing through the porous electrode, whereas how the electrode thickness impacts the microbial inactivation efficiency is not clear and needs to be studied systematically.

3.4.1 Investigation of the Effect of Electrode Pore Size on Bacterial Disinfection

The effect of electrode pore size on the disinfection performance of nanowire-assisted electroporation was analyzed with a controlled external voltage of 1 V. The results are shown in Fig. 3.10. When with the same HRT, the disinfection efficiency decreases with the increase of electrode pore size. When the HRT was 10 s, the water flow rate was the slowest in the experimental group and the microbes stayed in the electrode for the longest time. At this time, electroporation achieved complete inactivation of microbes in water (log removal rate >6.0 log) with the electrode pore size <700 μm. When lowered the HRT to 7 and 5 s, the maximum electrode pore size required for complete inactivation in water was 500 and 200 μm, respectively. When the HRT was further reduced to 3 s, the efficiency of microbial inactivation by electroporation was not significant (2.7 log) even if the electrode pore size was reduced to 100 μm due to the fast flow rate.

3.4.2 Investigation of the Effect of Electrode Thickness on Bacterial Disinfection

The effect of electrode thickness on nanowire-assisted electroporation disinfection was analyzed using copper foam electrodes with different electrode thicknesses (1,

Fig. 3.10 Effect of electrode pore size on electroporation disinfection performance

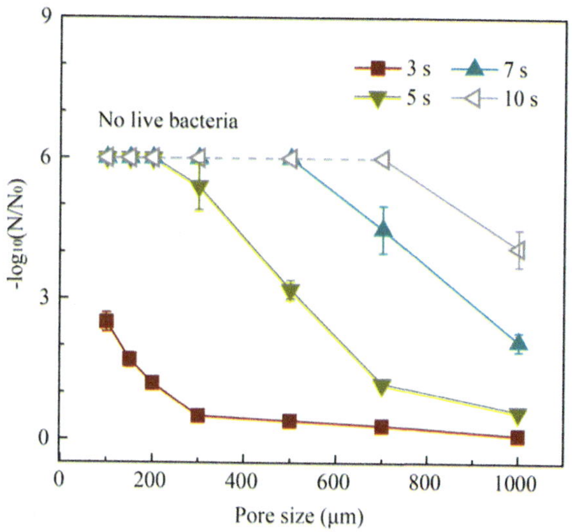

2, 5, and 10 mm). The applied voltage was fixed at 1 V and HRT was fixed at 5 and 7 s. As the results shown in Fig. 3.11, for the same HRT, the disinfection efficiency increased with electrode thickness, while the increase was not significant. When fixed HRT at 5 s and pore size at 100 μm, the water flow was the fastest and the device achieved complete microbial inactivation at various electrode thicknesses from 1 to 10 mm. As increasing the electrode pore size to 200 μm, 4.2 log microbial inactivation was achieved when the electrode thickness was 1 mm. Increasing the electrode thickness to 2 mm achieved complete disinfection (log removal rate >6.0 log). When the electrode pore size is increased to 500 μm, only 1.2 log microbial removal is achieved at the electrode thickness of 1 mm. When the electrode thickness increased to 10 mm, 3.7 log microbial removal can be achieved. At this point, a tenfold increase in electrode thickness resulted in only a twofold increase in disinfection efficiency. When the electrode pore size was 1000 μm, the microbial inactivation efficiency is lower (<1.0 log) for various electrode thicknesses.

When fixed the HRT at 7 s, the water flew through the electrode slow. The copper foam electrodes with pore size of 100 and 200 μm achieved complete disinfection at various electrode thicknesses from 1 to 10 mm. When the pore size was 500 μm, the electrode with thickness of 1 mm achieved ~3.0 log bacterial inactivation. As increasing the electrode thickness to 2 mm, the complete bacterial inactivation was achieved. When the electrode pore size was 1000 μm, only 0.6 log microbial inactivation was achieved with the 1 mm electrode, and 2.5 log microbial inactivation was achieved with the 10 mm electrode. A tenfold increase in electrode thickness resulted in only a fourfold increase in disinfection efficiency.

In summary, compared to the significant effect of electrode pore size on nanowire-assisted electroporation disinfection, the effect of electrode thickness is less dominating. Thicker electrodes do not lead to significant disinfection performance

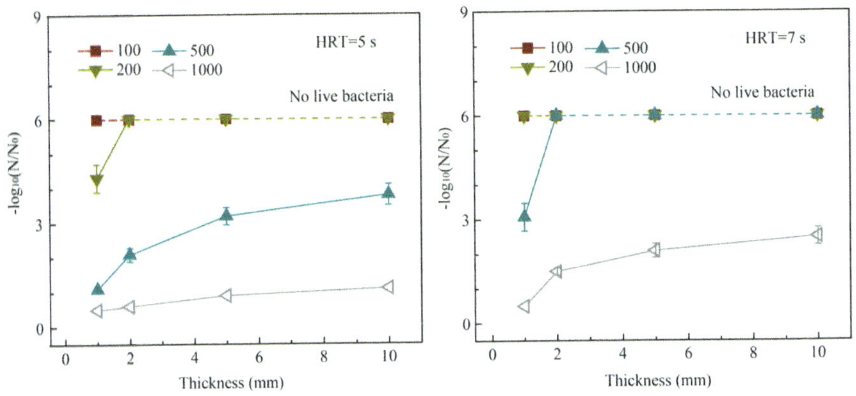

Fig. 3.11 Effect of electrode thickness on electroporation disinfection performance

improvement. Therefore, when designing electrodes for nanowire-assisted electroporation disinfection, thin electrodes (2–5 mm) can be used to ensure disinfection efficiency while reducing fabrication costs. Appropriate electrode pore size needs to be selected according to actual water quality conditions and treatment requirements.

3.5 Investigation on the Effect of Electrode Arrangement on Bacterial Disinfection

After clarifying the effect of nanowire morphology and electrode structure on nanowire-assisted electroporation disinfection, the study further analyzed the effect of electrode arrangement on bacterial disinfection. Considering that natural water bodies usually have a certain ionic strength (about 10 mM) neutral pH, when existing in natural water, the bacterial surface will be negatively charged because the bacterial surface isoelectric point is between 3 and 5. The negatively charged bacteria are more likely to be electrostatically attracted closer to the positive electrode. Therefore, during treating actual water samples, the positive and negative reactor contributions may be different and their contribution to disinfection needs to be studied systematically.

3.5.1 Contribution of Positive and Negative Electrodes to Microbial Inactivation During Nanowire-assisted Electroporation Disinfection

To investigate the contribution of positive and negative electrodes to bacterial inactivation in DI water and 10 mM NaCl solution (simulating the ionic strength of natural water), experiments were set up using effective and control electrodes. The effective electrode was prepared by heating a copper foam electrode (pore size of 500 μm and thickness of 5 mm) to 400 °C at a heating rate of 2.2 °C/min and keeping it at 400 °C for 2 h. The electrode prepared under this condition has high efficiency of electroporation microbial inactivation. The control electrode was prepared by heating the copper foam electrode with the same operating conditions to 600 °C at a heating rate of 20 °C/min and keeping the heating at 600 °C for 2 h. By this method, the electrode surface was only covered with a layer of copper oxide but no nanowire structure, which did not contribute to the electroporation disinfection.

The positive and negative electrodes were set as the effective electrode respectively, while the corresponding electrode was the control electrode, and *E. coli* was sed into DI water and 10 mM NaCl solution respectively. The disinfection performance was tested at an external voltage of 1 V and an HRT of 7 s. As shown in Fig. 3.12, in DI water, the positive and negative electrodes had similar microbial disinfection efficiencies of 4.5 log and 4.4 log, respectively. This indicated that the positive and negative electrodes contributed equally to the electroporation disinfection in the DI water system. However, in the 10 mM NaCl solution, the positive electrode had a higher efficiency (5.5 log) than the negative electrode (3.1 log), which indicates that the positive and negative electrodes contribute differently to the electroporation disinfection when treating actual water samples. The positive electrode was confirmed to contribute more than the negative electrode.

To analyze the mechanism of the greater disinfection contribution of the positive electrode in actual water, the bacteria were sed into NaCl solutions with different pH values while adjusting the final solution ionic strength at 10 mM. The surface charge distribution of the bacteria in the solutions was measured at different pH values. The external voltage was set at 1 V and the HRT was 7 s. The microbial

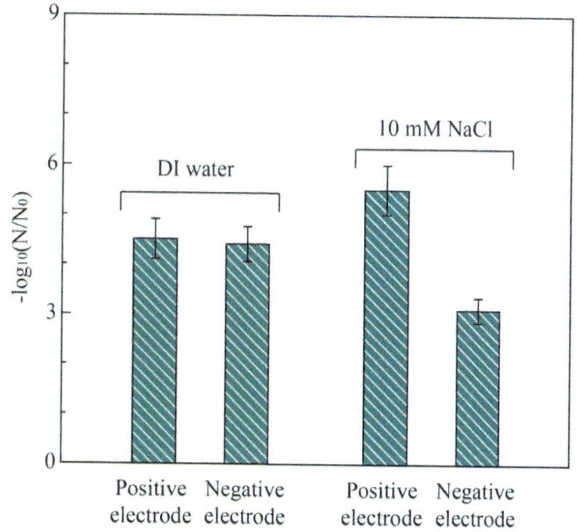

Fig. 3.12 Contribution of positive and negative electrodes on electroporation disinfection performance

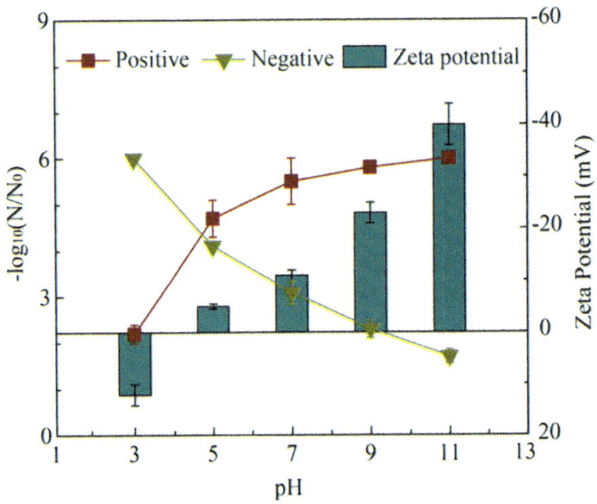

Fig. 3.13 Investigation on electrode and bacterial surface charge during electroporation disinfection

disinfection performance at different pH solutions was measured for positive and negative electrodes, respectively. As shown in Fig. 3.13, the Zeta potential on the bacterial surface decreases with the increase of pH. At pH 3, the bacterial surface had a positive charge (12 mV), while when the pH increased to 5, the bacterial surface has a negative charge (−5 mV). When the positive electrode is the effective electrode, there is a strong correlation between the disinfection efficiency of the electrode and the charge on the bacterial surface. The more negative charge on the bacterial surface, the higher the disinfection efficiency of the positive electrode. When the negative electrode is the effective electrode, the disinfection efficiency is the opposite of the surface charge of the bacteria, i.e., the more negative charges on the surface of the bacteria, the lower the disinfection efficiency of the negative electrode. Above results indicated that the bacterial charge is the main influencing factor for the different disinfection efficiency of the positive and negative electrodes. Under pH-neutral conditions, bacteria are negatively charged and attracted at the positive electrode, so the process for bacteria approaching the electrode surface was accelerated, leading to a higher disinfection efficiency.

3.5.2 Reactor Optimization to Enhance Electroporation Disinfection Efficiency

It is clear that the positive electrode contributes more to the electroporation disinfection in the actual treatment process. Therefore, the design of the reactor needs to be optimized for both positive and negative electrodes to achieve the maximum utilization of the electrodes. When optimizing the electrode parameters includes electrode pore size and electrode thickness, it is more significant to adjust the pore size to improve the disinfection efficiency than to adjust the electrode thickness. Therefore, in this study, the positive and negative electrodes were set at the same thickness (5 mm) and the effect of different combinations of positive and negative with various pore sizes on disinfection efficiency was investigated. As shown in Fig. 3.14, when the HRT was 5 s and the pore size of both positive and negative electrodes was 200 μm, the electroporation disinfection device achieved complete disinfection (log removal rate >6.0 log). When the negative pore size increased to 500 μm and the positive pore size was kept at 200 μm, the electroporation disinfection device still achieved complete disinfection. Although increasing the pore size is supposed to lower the disinfection efficiency, the negative electrode contributed less to disinfection, the lowered disinfection efficiency at the negative electrode had a limited impact on the overall disinfection efficiency. In fact, ensuring the positive electrode disinfection efficiency is essential for the overall reactor disinfection efficiency due to the larger contribution. When reversing the electrode setup and fixing the negative pore size at 200 μm and the positive pore size at 200 μm, the disinfection efficiency at this point was lowered and similar to that at 500 μm for both positive and negative apertures (3.5 log).

A similar experimental phenomenon can be achieved when fixed the HRT at 7 s with different electrode pore sizes. When both positive and negative electrode pore sizes were 500 μm, the electroporation disinfection device achieved complete disinfection. By increasing the negative electrode pore size to 700 μm and keeping the positive electrode pore size at 500 μm, complete disinfection in water is still achieved. However, when the electrode pore sizes were reversed (i.e., the positive electrode pore size 700 μm and the negative electrode pore size 500 μm), the disinfection efficiency was significantly reduced to 4.5 log. The lowered disinfection performance was similar to that of the positive and negative electrodes with the same pore size at 700 μm (4.7 log). In summary, since in the process of treating actual water bodies, the ideal method for optimizing the electrode is to reduce the positive electrode pore size to strengthen the positive treatment efficiency, thus ensuring the overall efficiency.

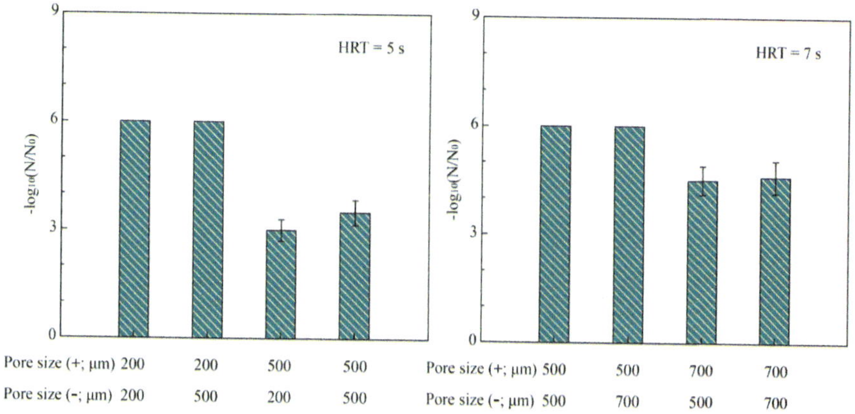

Fig. 3.14 Reactor optimization to enhance electroporation disinfection efficiency

The negative electrode pore size can appropriately increase, which not only has no significant effect on the overall efficiency but also reduces the electrode fabrication cost.

3.6 Summary

(1) To master the method of controlling nanowire morphology and analyze the impact of nanowire morphology on electroporation disinfection. The nanowire diameter and aspect ratio are the main factors affecting electroporation disinfection. To achieve efficient nanowire-assisted electroporation disinfection, the nanowire diameter on the electrode surface needs to be at the nanometer level and the aspect ratio needs to be greater than 40. A nanowire structure with a high aspect ratio can effectively enhance the electric field near the electrode surface and achieve efficient electroporation disinfection.

(2) Analysis of the influence of electrode parameters on electroporation disinfection. Electrode pore size and electrode thickness are both key factors affecting electroporation disinfection, with electrode pore size having a more significant effect on electroporation disinfection efficiency. As the electrode pore size decreased, the electroporation disinfection efficiency increased significantly, while as the electrode thickness increased, the electroporation disinfection improved not significantly.

(3) Analysis of positive and negative electrode contributions and optimization of electrode setup. Positive electrodes have a greater disinfection contribution when treating actual water bodies. It is clarified that the electrode combination needs to be optimized by reducing the positive electrode pore size to strengthen the positive electrode treatment efficiency, thus guaranteeing the overall efficiency. Meanwhile, appropriately increasing the negative electrode pore size can save the electrode fabrication cost without significantly affecting the overall treatment efficiency.

Chapter 4
Fabrication of High-Durability Nanowire-Modified Electrodes and Investigation of Their Microbial Disinfection Performance

4.1 Research Background

When nanowires with ideal morphology are grown on the surface of the porous electrode, a strong electric field can generate at the tip area of the nanowires even with a low external voltage (1 V). When bacteria in water flow through the porous electrode, they can quickly approach the surface of the porous electrode and enter the area with a strong electric field, which leads to electroporation disinfection. Since the nanowire structure is a slender structure with a small diameter and relatively large length, it is easy to be damaged when the water flows rapidly through the nanowire. Meanwhile, when there is a certain concentration of bacteria in the water flow, the bacteria with ~1 μm in diameter will continuously hit the nanowires with the water flow, which will aggravate the damage of the nanowire structure. In addition, during the treatment process, a certain voltage of direct current needs to be applied to both the positive and negative electrodes, so it is inevitable to cause redox reactions on the electrode surface, resulting in electrode corrosion loss. Both physical detachment and chemical corrosion will lead to nanowire-modified electrode damage, which not only affects the electrode lifetime but also causes new ecological risks from the released heavy metal ions. Therefore, fabricating high-durability and stable nanowire-modified porous electrodes and improving electrode stability are key issues for the practical application of nanowire-assisted electroporation disinfection, and are also technical challenges for researchers at present.

The objective of this chapter is to fabricate the high-durability and stable nanowires on the surface of porous electrodes. Copper phosphide (Cu_3P) nanowires with good physical strength and chemical stability can be uniformly grown on copper foam electrodes. An electroporation disinfection device can be built using the Cu_3P nanowire-modified copper foam electrodes, which can improve the continuous operating time of electrodes, reduce the concentration of the released copper ions in the effluent, and ensure water safety. The disinfection device was also used to evaluate the inactivation performance of typical model microbes in the reclaimed water.

Z. Huo, *Nanowire-assisted Flow-through Electrode Enabling Electroporation Disinfection of Reclaimed Water*, Springer Theses,
https://doi.org/10.1007/978-981-99-4502-3_4

4.2 Experimental Materials and Methods

4.2.1 Experimental Reagents

Sodium hypophosphite was purchased from Sigma, USA, Graphite paper was purchased from Beijing Jinglong Special Carbon Company, and the rest of the reagents were the same as in Sect. 2.2.1.

4.2.2 Fabrication of Cu_3P Nanowire-Modified Copper Foam Electrode

The copper foam (the same in Sect. 2.2.1) was selected as the porous electrode to build the nanowire-assisted electroporation disinfection device. The copper foam was cut into a cylindrical electrode with a diameter of 10 mm and a thickness of 2 mm. The pore size of the copper foam electrode was 200 μm. Cu_3P nanowires were grown *in-situ* on the copper foam surface using a two-step method.

(1) Fabrication of copper hydroxide ($Cu(OH)_2$) nanowires by electrochemical oxidation

The prepared copper foam electrode was washed first with 1 M HCl (Sigma) and then subsequently with DI water to remove any surface impurities. As shown in Fig. 4.1, the cleaned copper foam electrode was connected to the positive terminal of the power supply (Keithley, 2400, USA) and the graphite paper (2.0 × 3.5 cm) as the inert electrode was connected to the negative terminal of the power supply. The positive and negative electrodes were submerged in NaOH (1.5 M), and the power supply was turned on for the anodic oxidation reaction with a controlled current of 10 mA. After 20 min of anodic oxidation, wash the copper foam using DI water repeatedly and $Cu(OH)_2$ nanowires grew *in-situ* on the surface of copper foam.

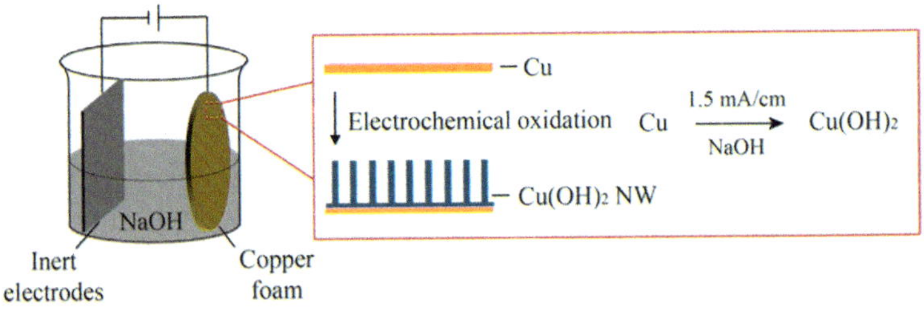

Fig. 4.1 Fabrication of $Cu(OH)_2$ nanowires using electrochemical oxidation

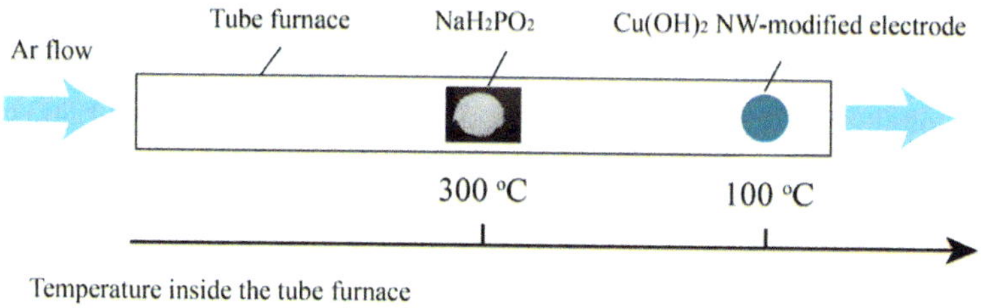

Fig. 4.2 Fabrication of Cu_3P nanowires by phosphidation process

(2) Fabrication of Cu_3P nanowires by phosphidation

As shown in Fig. 4.2, an excess of sodium hypophosphite (5 mg) was placed in the center of the tube furnace (Tianjin Zhonghuan Electric Furnace Co., Ltd., SK-G06123K), and the prepared $Cu(OH)_2$ nanowires-modified copper foam electrodes were placed at the downside of the tube furnace in the direction of airflow. After flushing with Ar for 15 min, the center of the furnace was elevated to 300 °C and the electrode with $Cu(OH)_2$ nanowires was placed downstream of the furnace where the temperature was 120 °C. After 90 min, the furnace cooled down naturally to room temperature under an Ar atmosphere and the Cu_3P nanowire-modified copper foam electrode was obtained.

4.2.3 Construction of Nanowire-assisted Electroporation Disinfection Devices

The disinfection device used for the study was based on the device used in Sect. 2.2.2.

4.2.4 Characterization and Elemental Analysis of Nanowire-Modified Electrode

(1) Morphological characterization

The nanowire modified copper foam electrodes were characterized by SEM in the same way as in Sect. 2.2.3. The scanning transmission electron microscopy (STEM) sample preparation method is the same as the TEM sample preparation method (same as Sect. 2.2.3).

(2) Elemental analysis

The point analysis and surface analysis of the elemental distribution of Cu_3P nanowires were performed by using an energy dispersive X-ray spectrometer (EDX) with SEM and STEM.

The mass ratio of Cu to P elements of Cu_3P nanowires was determined by ICP-MS. The prepared Cu_3P nanowire-modified copper foam electrode was placed in 10 mL of ethanol solution and sonicated for 20 min. Centrifuge the solution at 14,500 rpm for 10 min at room temperature (Hitachi, RX2 series) and discard the supernatant. The deposit was then dissolved with 1.0 mL of nitric acid (1 M) and the concentrations of copper and phosphorus ions were finally determined by ICP-MS (Thermo Scientific, XSERIES 2).

(3) Structural characterization

TEM was used to structurally characterize the Cu_3P nanowire-modified copper foam electrode, and the characterization method was the same as in Sect. 2.2.3. X-ray spectroscopy (XRD, PANalytical, Alpha 1 MPD) was used for the structural characterization of the Cu_3P nanowire-modified copper foam electrode. The scanning angles were from 12.5 to 35°.

(4) Electrochemical characterization

The electrochemical properties of the electrodes were determined using cyclic voltammetry (CV). A three-electrode system was constructed using an electrochemical workstation (BioLogic, VMP3) for the measurements, using an Ag/AgCl electrode as the reference electrode and a graphite paper electrode (2.0 × 3.5 cm) as the counter electrode. The measurements were carried out in a 0.1 M phosphate buffer system. A voltage of -0.7 to 0.5 V (vs. Ag/AgCl electrode) was applied to the working electrode during the measurement and the voltage scan rate was controlled to be 10 mV/s.

4.2.5 Microbes and Water Samples Used in Experiments

Enterobacter hormaechei (*E. hormaechei*; ATCC 700,323) was purchased from Bei Na Biological Company, China. After purchase, the bacteria were stored in a refrigerator at −80 °C.

E. hormaechei suspension: except that the liquid medium for culture was Tryptic Soy Broth medium (TSB; see Table 4.1 for composition), the rest of the configuration steps were the same as those for *E. coli* suspension in Sect. 2.2.4.

Table 4.1 Composition of the tryptic soy broth used in the study

Composition	Conc. (g/L)
Casein peptone	17
Soya peptone	3
K_2HPO_4	2.5
Glucose	2.5
NaCl	5

The rest of the strain culture and water sample configuration method with Sect. 2.2.4.

4.2.6 Cu_3P Nanowire-assisted Electroporation for Microbial Disinfection

The experimental method is the same as in Sect. 2.2.5.

4.2.7 *Analysis of Microbial Inactivation Mechanisms*

(1) Analysis of high-frequency alternating current (AC) disinfection efficiency

When an external voltage is applied to water, a variety of physical and chemical processes, including electrochemical and electroporation, may occur. Different physical and chemical processes have specific minimum response times. Electrochemistry is usually considered to fail when the time of applied voltage is less than 1 ms. In the case of electroporation, the electrons inside and outside the cell membrane are redistributed in response to an external voltage, generating a transmembrane potential, a process similar to charging a capacitor. When the external voltage is applied for less than 1 μs, the time is too short for the electrons inside and outside the cell membrane to be redistributed, making it difficult for electroporation to occur.

Different application times of external voltage can be controlled using different frequencies of AC. The higher the AC frequency, the shorter the time of applied voltage in a single cycle. By applying different frequencies of AC, electrochemistry and electroporation can be effectively excluded from the treatment process. In this study, a high-frequency signal generator (Beijing Puyuan, DG5000) was used as the experimental power source to apply AC voltages at frequencies from 10 to 3.5×10^8 Hz (peak voltage of 1 V) during the disinfection process and to simultaneously record the current intensity in the circuit.

(2) Microbial Morphology analysis before and after electroporation disinfection

The experimental method is the same as in Sect. 2.2.9.

(3) Bacterial staining analysis

The experimental method is the same as in Sect. 2.2.10 except that the stain used is PI only.

4.2.8 Analysis of the Disinfection Efficiency Using Nanowire-Modified Electrodes for Long-Term Operation

The prepared electrodes were applied in an electroporation disinfection device with a fixed applied voltage (1 V) and HRT (5 s). The inactivation efficiency of *E. coli* during long-term operation was evaluated.

4.2.9 Analysis of the Loss Mechanism of Electrode During Long-Term Operation

(1) Copper loss analysis during long-term operation

During long-term operation, copper loss from copper foam electrodes was analyzed. The copper loss in the electrode mainly includes the loss of copper ions due to electrochemical dissolution and the loss of copper particles due to water scouring. The measurement methods are as follows.

(1) Collect 1 mL of water sample after a certain time of nanowire-assisted electroporation disinfection and add 1 mL of HNO_3 (2 M) to make the final HNO_3 concentration of 1 M. The concentration of copper ions in the sample was detected by ICP-MS (Thermo Scientific, XSERIES 2), which is the total copper loss concentration, and recorded as C_T.
(2) Another 1 mL of water sample was collected, centrifuged at 14,500 rpm for 10 min at room temperature (Hitachi, RX2 series), and the concentration of copper ions in the supernatant was measured by ICP-MS (Thermo Scientific, XSERIES 2), which is the concentration of copper ions dissolved at the electrode due to the electrochemical reaction, and is recorded as C_{dis}.
(3) The concentration of copper particles lost from the copper electrode due to water scouring is C_{det}, calculated as $C_{det} = C_T - C_{dis}$.

(2) Electrode morphology analysis during long-term operation.

The Cu_3P nanowire-modified copper foam electrode was taken at different operating times and placed in a drying dish for 12 h. The morphological analysis of the dried electrode was performed by SEM, and the experimental method was the same as that in Sect. 2.2.3.

(3) Analysis of electrode loss mechanism during long-term operation.

Cu_3P nanowire-modified copper foam electrodes were taken at different operating times and dried in a drying dish for 12 h. XRD was used to structurally characterize the substances of the electrodes after a long-term operation. The experimental method was the same as in Sect. 4.2.4.

X-ray photoelectron spectroscopy (XPS, Kratos Analytical, K-alpha) was used to characterize the surface elemental chemistry of the Cu_3P nanowire-modified copper foam electrode. A single-crystal aluminum target was used as the X-ray source under a high vacuum (<10^{-8} Torr), and a 20 × 20 μm area on the material surface was selected for the surface analysis.

4.3 Fabrication and Characterization of Cu_3P Nanowire-Modified Electrodes

4.3.1 Fabrication of Cu_3P Nanowire-Modified Electrodes

As shown in Fig. 4.3, Cu3P nanowires were grown *in-situ* on the surface of copper foam using a two-step process, and the electrode color changed from brown (copper foam) to green ($Cu(OH)_2$ nanowire-modified copper foam electrode) and then to black (Cu_3P nanowire-modified copper foam electrode) in sequence.

After the two-step fabrication, Cu_3P nanowires uniformly grew on the surface of the copper foam electrode, and their morphology was shown in Fig. 4.4. The length of Cu_3P nanowires was ~20 μm and the diameter was ~400 μm, and the aspect ratio was calculated to be 50. Cu_3P nanowires grew vertically and uniformly on the surface of the electrode substrate with a stable and robust structure. According to the high magnification SEM images, the nanowires were tightly bonded to the electrode substrate, which ensured that they were not easily broken by the water flow. The morphological analysis of the nanowire tip showed that the Cu_3P nanowire tip structure was robust and had high physical stability, which ensured the strength of the nanowire structure.

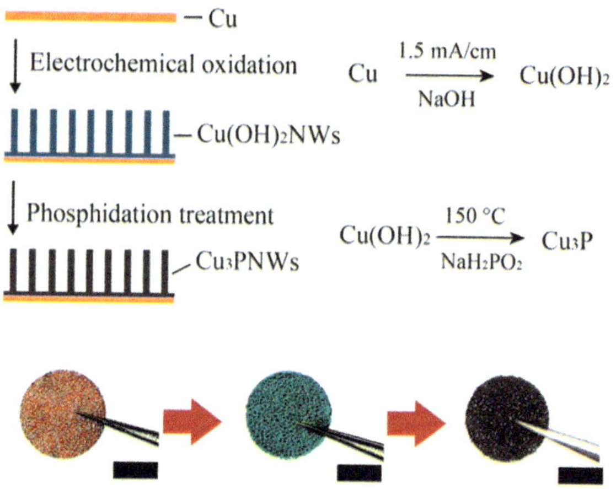

Fig. 4.3 Schematic of fabricating the Cu_3P nanowire-modified electrode. The scale bar is 1 mm

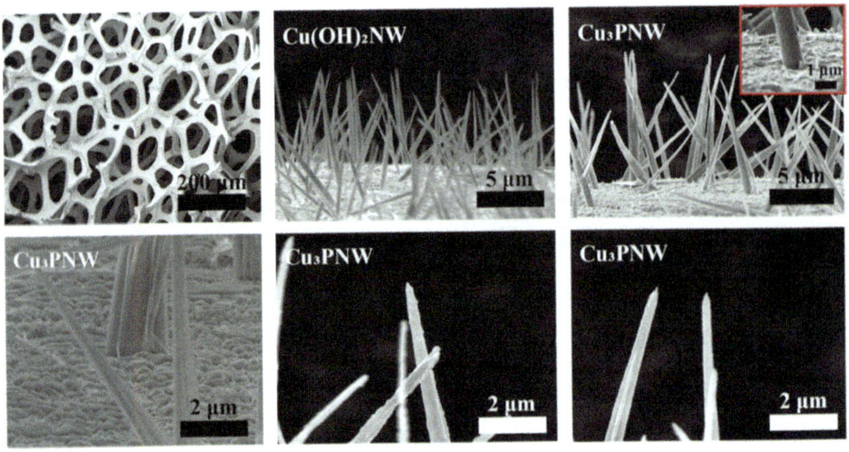

Fig. 4.4 Morphological analysis of the Cu_3P nanowire-modified electrode

4.3.2 Characterization of Cu_3P Nanowire-Modified Electrodes

As the TEM characterization results shown in Fig. 4.5, the diameter of the Cu_3P nanowire was ~400 nm, and the surface of the nanowire was uneven at some locations, which was due to the polycrystalline material of the synthesized material. The crystalline surface spacing of the material was 2.00 Å by HRTEM analysis, which matched with the specific crystalline surface spacing of Cu_3P and proved that the synthesized material was Cu_3P.

In addition, STEM was applied for the morphology investigation of single Cu_3P nanowires. As shown in Fig. 4.6, the surface scan and projection analysis by STEM indicated that the synthesized Cu_3P nanowires had a uniform surface and internal structure. The elemental surface analysis by STEM with EDX configuration proved that the synthesized material was mainly composed of two elements, Cu and P.

Fig. 4.5 TEM characterization of Cu_3P nanowires

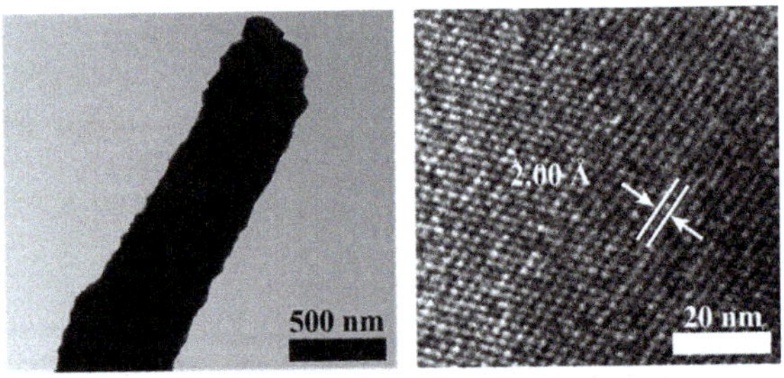

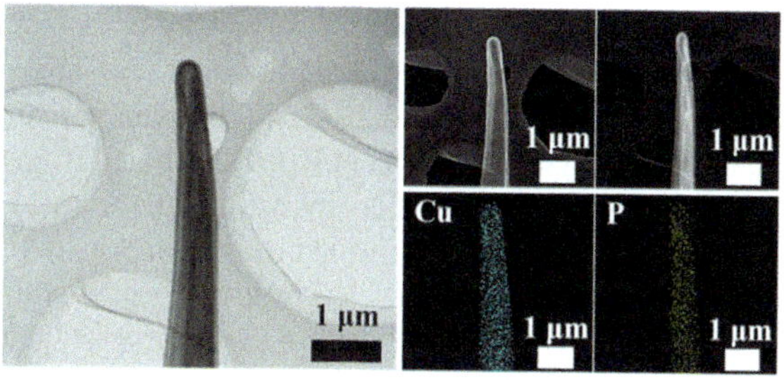

Fig. 4.6 STEM morphological characterization and EDX elemental analysis of Cu_3P nanowires

The elemental quantification of a point-analysis on the Cu_3P nanowire was performed using an EDX detector in SEM configuration. At the location of the EDX sampling point shown (Fig. 4.7), the nanowire material contained only two elements, Cu and P, and the ratio of atoms of Cu to P elements was about 3:1. This indicated that the synthesized material was Cu_3P.

Considering that there may be some system errors in the measurement process when using the EDX for elemental analysis of materials, the above-mentioned EDX point elemental analysis is usually only used as one supporting method for quantitative analysis. To further perform accurate quantitative analysis of the elements contained in nanowires, the nanowire materials grown *in-situ* on the electrode surface

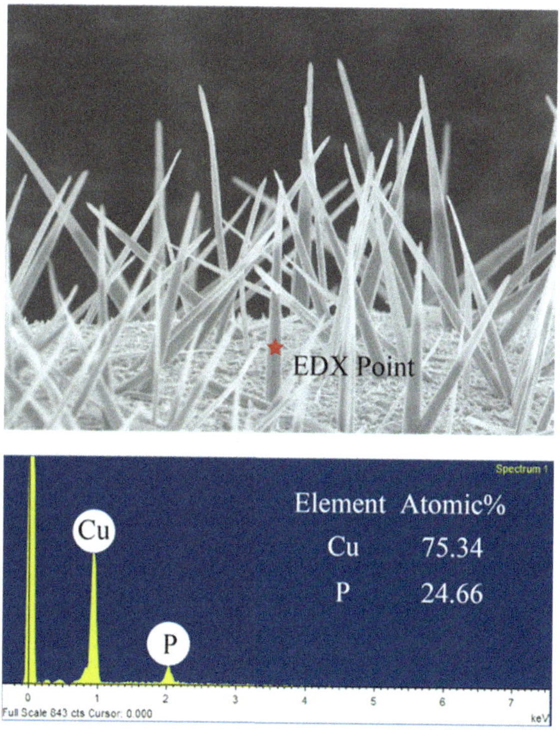

Fig. 4.7 EDX point elemental analysis of Cu_3P nanowires

were stripped from the electrode surface by ultrasonication. The exfoliated nanowire material was collected, dissolved in nitric acid, and the elemental proportions were accurately determined using ICP-MS for quantitative analysis of the contained materials. As shown in Fig. 4.8, nanowires exfoliated from the electrode surface contained both Cu and P elements, where the ratio of the atomic number of Cu to P elements was about 3:1. This further confirmed that the fabricated nanowire material was Cu_3P.

The crystal structure of Cu_3P was characterized by XRD. As shown in Fig. 4.9, besides the two strong diffraction peaks at 43.4° and 50.6° from the Cu substrate (JCPDS file no. 04–0836), the other peaks can be assigned to the Cu_3P phase, confirming the successful synthesis of Cu_3P nanowires on the electrode surface. The diffraction pattern for the Cu_3P nanowire-modified copper foam electrode exhibits six broad peaks at 36.0°, 39.1°, 41.6°, 45.1°, 46.2°, and 47.3°, corresponding to (112), (202), (211), (300), (113), and (212) of Cu_3P phase (JCPDS file no. 71–2261), respectively.

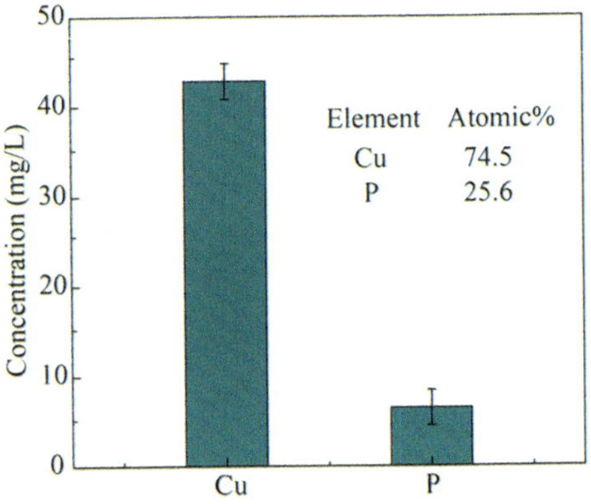

Fig. 4.8 Elemental analysis of Cu_3P nanowires by ICP-MS

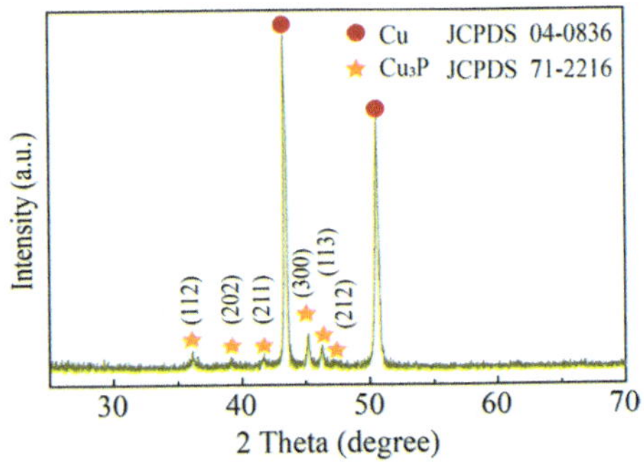

Fig. 4.9 XRD pattern of the Cu_3P nanowire-modified copper foam electrode

4.4 Disinfection Efficiency and Mechanism of Nanowire-assisted Electroporation Using Cu_3P Nanowire-Modified Electrodes

4.4.1 *Disinfection Efficiency of Nanowire-assisted Electroporation Using Cu_3P Nanowire-Modified Electrodes*

The Cu_3P nanowire-modified copper foam electrode was installed in the electroporation reactor and the microbial inactivation efficiency was measured under different operating conditions by controlling the external voltage and the flow rate of the water sample. As shown in Fig. 4.10, The Cu_3P nanowire-modified copper foam electrode achieved efficient inactivation of *E. coli* at low voltage, and the microbial inactivation efficiency increased with the increase of external voltage and HRT. At an external voltage of 1 V, the Cu_3P nanowire-modified copper foam electrode achieved completely disinfection at an HRT of 5 s (removal rate >6.0 log), and when the external voltage increased to 2 V and 3 V, the electrode achieved completely disinfection at an HRT of 4 s and 3 s, respectively. In addition, CuO nanowire-modified copper foam electrodes with the same electrode structure (pore size and thickness) can also achieve complete disinfection at an external voltage of 1 V and an HRT of 5 s. Thus, Cu_3P nanowire-modified electrodes and CuO nanowire-modified electrodes have similar microbial inactivation efficiency, and both can achieve efficient disinfection without chlorine at ultra-low voltage.

The study further analyzed the disinfection performance of Cu_3P nanowire-modified copper foam electrodes for different microbes. As shown in Fig. 4.10, for different bacteria, the Cu_3P nanowire-modified copper foam electrode achieved efficient inactivation. It is worth noting that the inactivation efficiency of Cu_3P

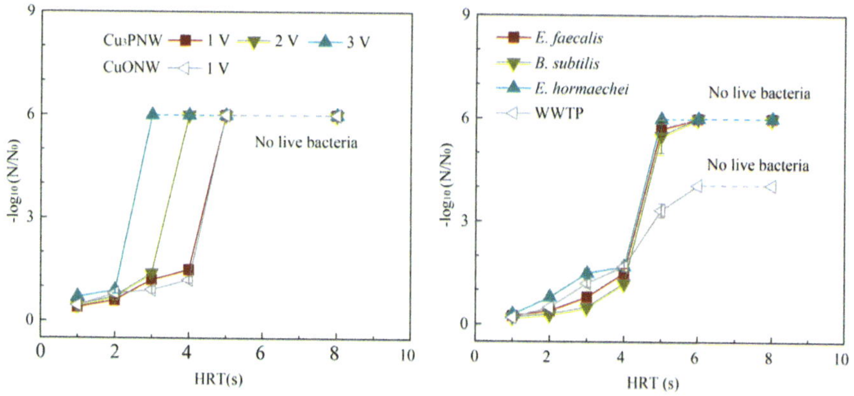

Fig. 4.10 Disinfection efficiency of nanowire-assisted electroporation using Cu_3P nanowire-modified electrodes

nanowire-modified copper foam electrode for Gram-positive bacteria (*E. faecalis* and *B. subtilis*) is slightly lower than that for Gram-negative bacteria (*E. hormaechei*), which may be attributed to the thick cell wall structure of Gram-positive bacteria showing higher resistance to electroporation. In addition, the disinfection performance of actual wastewater was evaluated using Cu_3P nanowire-modified copper foam electrodes. All the bacteria in actual wastewater were inactivated completely (removal rate >4.1 log) at an external voltage of 1 V and HRT of 6 s.

4.4.2 Disinfection Mechanisms of Nanowire-assisted Electroporation Using Cu_3P Nanowire-Modified Electrodes

The study analyzed the disinfection mechanism of the electroporation operation using Cu_3P nanowire-modified copper foam electrodes. When an external voltage is applied in water through positive and negative electrodes, there are various physical and chemical processes including electrochemical and electroporation. Each of these processes has a specific corresponding time. For electrochemistry, it is difficult to occur if the duration of the external voltage is less than 1 ms. In the case of electroporation, if the external voltage is less than 1 μs, no electroporation will happen. Therefore, the contribution of different factors to the disinfection can be examined by applying an AC to the electrodes and adjusting the duration of the external voltage in a single cycle by controlling the AC frequency.

A sinusoidal AC signal with a peak-to-peak voltage of 1 V was applied and the AC frequency (10 to 3.5×10^8 Hz) was adjusted to analyze the disinfection efficiency of Cu_3P nanowire-modified copper foam electrodes. As shown in Fig. 4.11, when the frequency was below 10^2 Hz, the Cu_3P nanowire-modified copper foam electrode achieved complete disinfection (log removal rate >6.0 log). At this time, there was a significant current (>200 μA) in the circuit, indicating that a certain electrochemical reaction was generated at this frequency. As the frequency increased to 10^6 Hz, the current in the circuit was significantly reduced to 10 μA, indicating that the electrochemical reaction was significantly limited while the Cu_3P nanowire-modified copper foam electrode still achieved complete disinfection. This indicated that electroporation instead of the electrochemical process was the dominant mechanism in the disinfection process. When the AC voltage frequency was further increased to 10^8 Hz, the efficiency of the Cu_3P nanowire-modified copper foam electrode for microbial disinfection was significantly eliminated. This indicated that when electroporation was excluded, the whole system had no contribution to microbial inactivation, which further demonstrating that electroporation was the most important disinfection mechanism in the disinfection process.

The study further analyzed the bacterial morphology before and after electroporation disinfection. A shown in Fig. 4.12, for the four typical model bacteria (Gram-negative bacteria: *E. coli* and *E. hormaechei*; Gram-positive bacteria: *B. subtilis* and

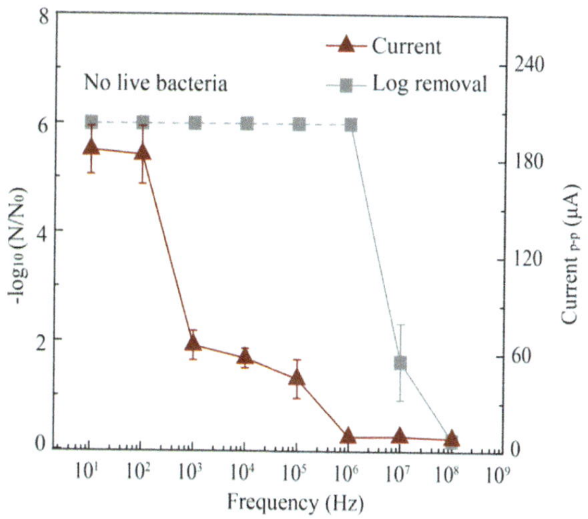

Fig. 4.11 Inactivation efficiency using Cu_3P nanowire-modified electrodes with different frequencies of AC signal (from 1 to 3.5×10^8 Hz) and the measured current intensity in the circuit

E. faecalis), the bacteria had a uniform, intact, and smooth cell membrane structure before the electroporation disinfection. However, after the electroporation operation (external voltage 1 V, HRT 6 s), the bacterial cell membrane showed small pores (<200 nm in diameter) on the cell membrane. The small pores with this morphology are typical electroporation pores. Therefore, the bacterial morphology analysis further proved that the disinfection mechanism was electroporation.

In addition, the study performed staining analysis of cells before and after electroporation disinfection to further analyze the mechanism of bacterial inactivation. As shown in Fig. 4.13, before the electroporation disinfection, the bacteria were not marked by PI staining, indicating that the bacteria had an intact cell membrane structure. After the electroporation operation (external voltage 1 V, HRT 6 s), the bacteria were observed to be labeled by PI dye at the corresponding locations under bright field, confirming that the cell membrane was broken after the electroporation

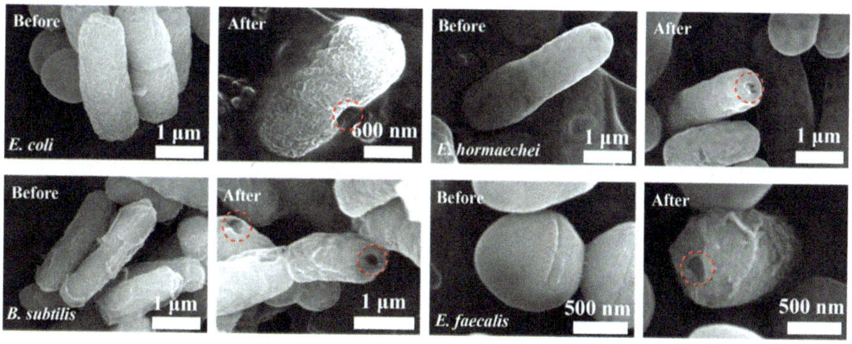

Fig. 4.12 Analysis of bacterial morphology before and after electroporation disinfection using Cu_3P nanowire-modified electrodes

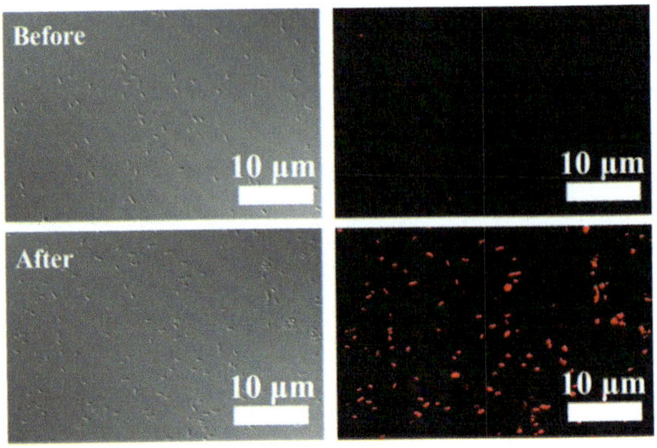

Fig. 4.13 Analysis of bacterial staining after Cu_3P nanowire-assisted electroporation disinfection

disinfection. This further proved the disinfection mechanism as electroporation. In summary, electroporation as a bacterial inactivation mechanism was demonstrated by high-frequency voltage analysis, cell morphology analysis, and bacterial staining analysis. It showed that Cu_3P nanowire-modified copper foam electrodes can be used for electroporation disinfection to achieve efficient microbial inactivation.

4.5 Long-Term Disinfection Performance and Electrode Loss Mechanism

4.5.1 Long-Term Disinfection Performance of Cu_3P Nanowire-Modified Electrodes

The core issue that needs to be considered during the practical application of nanowire-assisted electroporation disinfection is the stability of the nanowire-modified electrodes. The special structure of the nanomaterials leads to poor stability, which, together with the effect of water scouring and electrochemical corrosion, further aggravates the loss of electrodes and affects their lifetime. The loss of nanowire-modified electrodes during the operation process not only affects the electrode lifetime and leads to lower disinfection efficiency, but also the loss of heavy metals involves new health and ecological risks.

In the study, Cu_3P nanowires with a more robust physical structure and higher chemical stability were used to achieve electroporation disinfection. The microbial disinfection efficiency of Cu_3P nanowire-modified copper foam electrodes was analyzed during long-term operation. As shown in Fig. 4.14, the Cu_3P nanowire-modified copper foam electrode achieved complete disinfection (log removal rate >6.0) for tested microbes in water at an external voltage of 1 V and an HRT of 5 s. Under these operating conditions, the reactor can operate continuously for 12 h with no viable bacteria detected in the effluent. Nevertheless, under the same operating

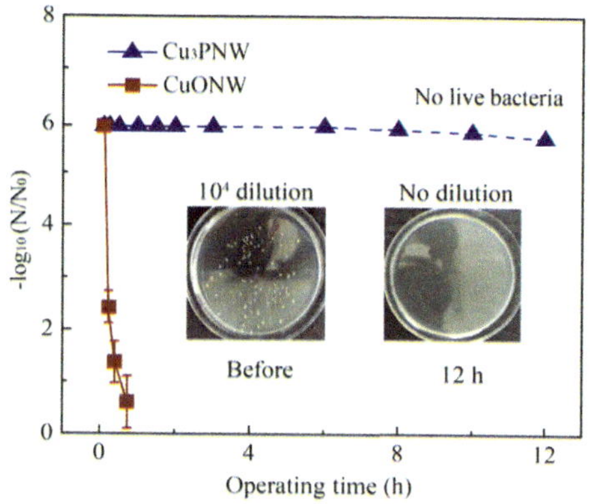

Fig. 4.14 Long-term disinfection efficiency of Cu_3P nanowire-modified copper foam electrodes

conditions, the CuO nanowire-modified copper foam electrode can only operate continuously for 15 min, and the disinfection efficiency of the CuO nanowire-modified copper foam electrode decreased significantly when the operating time exceeded 20 min. Therefore, the Cu_3P nanowire-modified electrode has good stability in the electroporation disinfection, and the electrode lifetime for complete disinfection is increased by nearly 50 times compared with that for CuO nanowire-modified electrodes.

4.5.2 Electrode Loss Phenomenon During the Long-Term Operation

The study further analyzed the electrode loss phenomenon during long-term operation when using Cu_3P nanowire-modified copper foam electrodes. Firstly, the copper loss during long-term operation was measured. As shown in Fig. 4.15, compared with CuO nanowire-modified copper foam electrode, Cu_3P nanowire-modified copper foam electrode showed significantly lower copper loss during long-term operation. For the CuO nanowire-modified electrode, the total concentration of the released copper electrode was 340 μg/L within the initial 0.2 h operation, which included 220 μg/L of copper ions released due to electrochemical dissolution and 120 μg/L of copper particles released due to water flow detachment. As the operating time increased, the concentration of copper ions remained stable, while the concentration of copper particles gradually decreased. It is highly possible that the concentration of copper particles detached by water flow gradually decreases due to the absence of nanowire structure on the electrode surface after a long period of operation. Compared with the CuO nanowire-modified electrode, which has more serious copper loss, the Cu_3P nanowire-modified copper foam electrode had significantly lower copper loss during the long-term operation. Within the initial 0.2 h operation, the total concentration of the released copper electrode was only 25 μg/L, which was only 7% of the CuO nanowire-modified electrode under the same operating conditions. With the

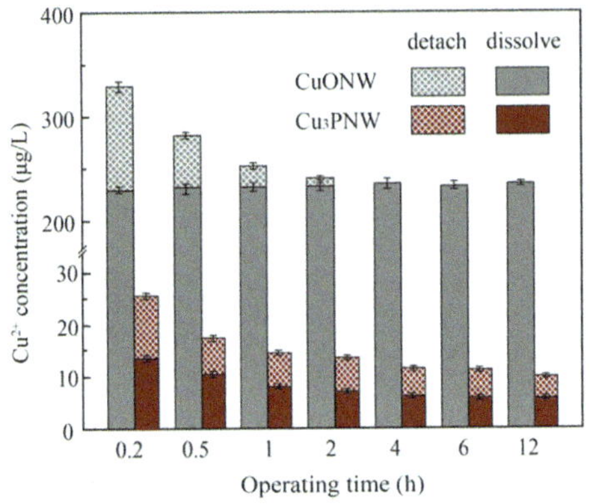

Fig. 4.15 Electrode loss phenomenon during long-term operation when using Cu_3P and CuO nanowire-modified copper foam electrodes

increase of the operating time, the copper loss from the electrode decreased, and after 12 h of continuous operation, the copper electrode loss concentration was 12 μg/L, including 7 μg/L of copper ions lost by electrochemical dissolution and 5 μg/L of copper particles lost by water flow detachment.

The Cu_3P nanowires can withstand more intense water detachment attributed to their robust structure, so the concentration of copper particles detached by water flow is significantly reduced. We further analyzed the electrochemical activity of Cu_3P and CuO nanowires using an electrochemical workstation. As shown in Fig. 4.16, the CuO nanowire-modified copper foam electrode had a significant current peak at a voltage of −0.4 versus 0.1 V (vs. Ag/AgCl), indicating that a redox reaction occurs at this voltage position. For the Cu_3P nanowire-modified copper foam electrode, there was also a certain current peak at a similar position, but the current intensity was significantly lower than that of the CuO nanowire-modified electrode. This indicated that the electrochemical activity of Cu_3P nanowire was significantly lower than that of CuO nanowire. This result also explained the trace loss of Cu ions due to dissolution during the long operation using Cu_3P nanowire-modified copper foam electrode.

4.5.3 Loss Mechanism of Cu_3P Nanowire-Modified Electrode

The mechanism of electrode loss during the long-term operation using Cu_3P nanowire-modified copper foam electrode was investigated and analyzed. Firstly, SEM was used to analyze the morphology of Cu_3P nanowire-modified electrode during long-term operation. As shown in Fig. 4.17, for both positive and negative electrodes, the Cu_3P nanowires maintained a good morphology with a length of 20 μm at the initial operation (0 h). As the operating time increased to 6 h, the length of Cu_3P nanowires at the positive electrode gradually decreased to 15 μm.

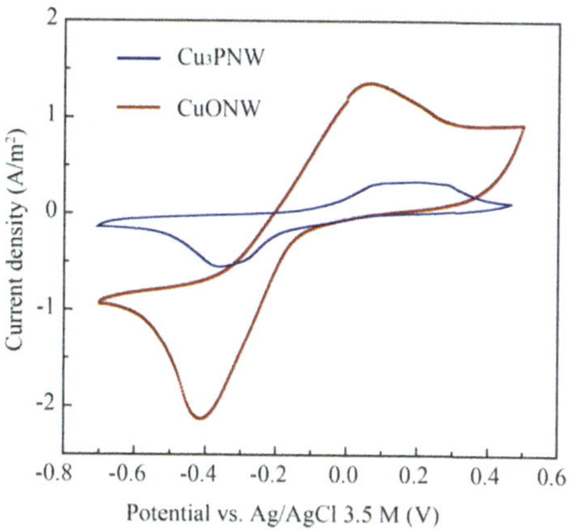

Fig. 4.16 Electrochemical activity analysis of Cu_3P and CuO nanowire-modified copper foam electrodes

The length of Cu_3P nanowires at the negative electrode not only decreased (average length 12 μm) but also formed crystal at its tip.

As the operating time increased to 12 h, the length of Cu3P nanowires at the positive electrode decreased to 5 μm, while the length of Cu3P nanowires at the negative electrode became shorter to ~3 μm with more crystallization aggregation (up to 1 μm in diameter) at the tip structure. The different morphologies of the positive and negative electrodes indicated that different redox reactions occurred at the different electrodes, leading to the loss and failure of the electrodes.

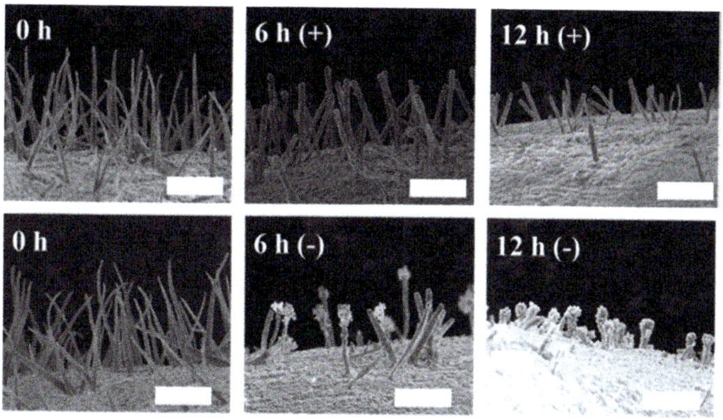

Fig. 4.17 Morphology analysis of Cu_3P nanowire-modified electrodes during long-term operation. Scale bars are 5 μm

Since the positive and negative electrodes undergo different electrochemical reactions, the study used the XRD technique to analyze the material structure changes of Cu_3P nanowire-modified copper foam electrodes during long-term operation. As shown in Fig. 4.18, during the initial operation (0 h), Cu_3P nanowire-modified electrode has a significant peak at 36.0°, which corresponds to the (112) crystal plane of Cu_3P material. As the operating time increased, the peak at 36.0° on the positive electrode gradually decreased, implying a gradual decrease of Cu_3P. At the same time, a new peak was generated at 31.0°, corresponding to the (021) crystal plane of $CuCO_3{\cdot}Cu(OH)_2$. this process can be explained by Eq. (4.1).

$$Cu_3P - 3e^- = 3Cu^{2+} + P^{3-} \quad (4.1)$$

Due to the oxidizing environment at the positive electrode, Cu_3P lost electrons and became Cu^{2+}, which created a region with high Cu^{2+} concentration on the electrode surface. This generated Cu^{2+} may react with CO_2 in the air to form $CuCO_3{\cdot}Cu(OH)_2$ during subsequent drying[152].

The peak at 36.0° on the negative electrode also gradually decreased, whereas a new at 34.0° generated, corresponding to the (002) crystal plane of $Cu(OH)_2$, which indicated the generation of new $Cu(OH)_2$ during the long-term operation[152]. Due to the localized reduction property near the negative electrode, the pH was alkaline and the Cu^{2+} dissolved from the positive electrode precipitated at the negative electrode. This process was also confirmed by SEM images showing the nanowire morphology with crystals on the tip. The electrochemical reaction occurring at the negative electrode can be explained by Eq. (4.2).

$$Cu_3P - 3e^- = 3Cu + P^{3-} \quad (4.2)$$

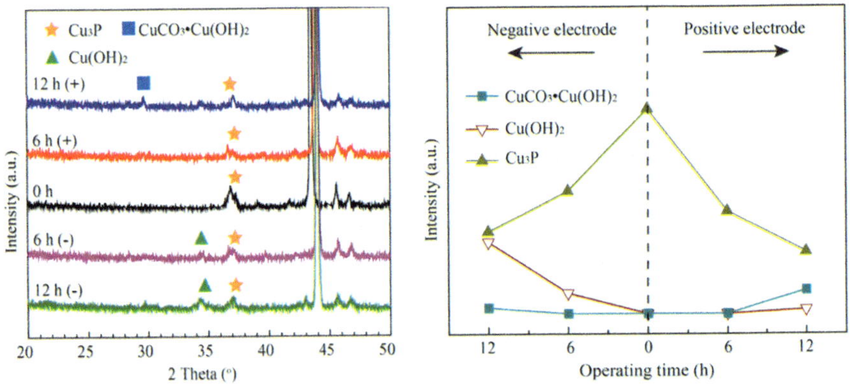

Fig. 4.18 Loss mechanism of Cu_3P nanowire-modified electrodes during long-term operation

Due to the reducing environment at the negative electrode, Cu_3P gained electrons and was reduced to small copper particles, which were released with the water flow. This phenomenon was consistent with the loss of Cu_3P during long-term operation: a small amount of copper particles can still be detected in the effluent during long-term operation at the negative electrode.

To further investigate the loss mechanism of Cu_3P nanowire-modified copper foam electrodes during long-term operation, XPS was used to analyze the surface elemental. As shown in Fig. 4.19, for the Cu_3P nanowire-modified electrode before operation, the copper element on its surface only showed a clear peak at the binding energy of 932.9 eV, which corresponded to the monovalent copper element Cu(I). This indicated that the surface of the electrode only contained Cu(I), i.e., Cu_3P, before operation. As the treatment time increased to 12 h, the 932.9 eV peaks at both positive and negative electrodes decreased and showed the satellite peak of Cu(II) at 945.5 eV[152]. This confirmed that compounds containing Cu(II) appeared on both positive and negative electrodes during the long-term operation, which was also consistent with the XRD analysis. As for elemental phosphorus, the peak shapes of elemental phosphorus on both positive and negative surfaces were similar before and after the operation, indicating that the elemental phosphorus was not involved in the reaction and was stable during the long-term operation.

In summary, based on the nanowire morphology analysis, electrode material analysis, and electrode surface element analysis, the loss mechanism of Cu_3P nanowire-modified copper foam electrode during the long-term operation can be summarized. As shown in Fig. 4.20, at the positive electrode, due to the oxidation reaction, some

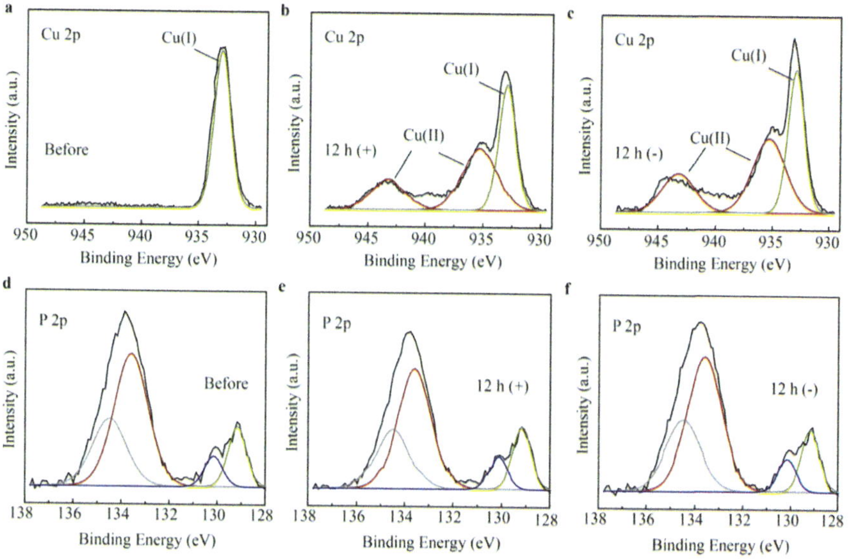

Fig. 4.19 Surface elemental analysis of Cu_3P nanowire-modified electrodes during long-term operation

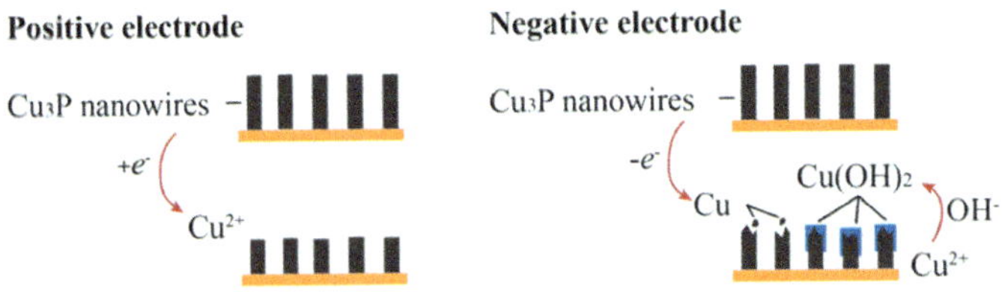

Fig. 4.20 Loss mechanism of Cu_3P nanowire-modified copper foam electrode during the long-term operation

Cu_3P was dissolved and released to the effluent as Cu^{2+}. At the negative electrode, the local pH is alkaline due to the reduction environment, leading to the deposition of Cu^{2+} released from the positive electrode in the form of $Cu(OH)_2$. At the same time, the local reduction reaction causes Cu_3P to be reduced to Cu particles and released to the effluent.

The clarification of the electrode loss mechanism can provide a theoretical basis for the subsequent enhancement of the electrode lifetime. Considering that the electrode loss is mainly due to the electrochemical reaction, in the subsequent study, the electrode surface modification or optimization of operating conditions can be applied to eliminate the electrochemical reaction for further extending the electrode lifetime. In addition, even if the electrode loss problem exists, due to the good physical and chemical stability of Cu_3P nanowires, Cu_3P nanowire-modified copper foam electrode can still achieve continuous disinfection for 12 h. It can be applied to small household devices including household filters, and at the same time, the treated water is safe due to the low copper concentration in the effluent (<25 ug/L). Therefore, Cu_3P nanowire modified-copper foam electrodes show great potential in the actual situation.

4.6 Summary

(1) The Cu_3P nanowire-modified copper foam electrode using a two-step fabrication method can achieve efficient water disinfection. Meanwhile, attributed to the strong physical and chemical stability of Cu_3P nanowires, a continuous complete disinfection for 12 h can be achieved, and the electrode lifetime time is nearly 50 times longer than that of CuO nanowire-modified copper foam electrode at the same operating conditions.

(2) It is revealed that Cu_3P nanowire-modified copper foam electrode shows high stability. During the long-term operation, the electrode loss is trace and slow with the copper concentration less than 25 ug/L, which is only 7% of the loss rate of the CuO nanowire-modified electrode.

(3) The electrode morphology analysis, electrode material structure analysis, and electrode surface element analysis reveal that the loss mechanism of Cu_3P nanowire-modified copper foam electrode is mainly due to electrochemical reaction during the long-term operation, which provides the theoretical basis for the further improvement of electrode lifetime.

Chapter 5
Surface Coating on Nanowire-Modified Electrode Lifetime Enhancement

5.1 Research Background

Cu_3P nanowire-modified copper foam electrodes with stable physical structure and high chemical stability can achieve efficient microbial disinfection during long-term operation. Mechanisms of electrode loss during the long-term operation of Cu_3P nanowire-modified electrode were confirmed to be the electrochemical redox reaction that occurred on the electrode surface and the electrode detachment by water flow due to the insufficient physical strength. Accordingly, the future research objective is to reduce or prevent the electrochemical reaction on the electrode surface and improve the nanowire strength to improve the electrode lifetime. Since the electrochemical reaction usually occurs at the electrode-solution interface, the use of coating technology to uniformly cover the electrode surface with a nanoscale film to form a protective layer can theoretically prevent the electrochemical reaction to improve the electrode lifetime. At the same time, a proper protective film can also enhance the strength of nanomaterials and resist water detaching during long-term operation. However, the above-mentioned nanowire-modified porous electrode surface modification methods have not been reported, and the uniform coating technology on the porous electrode surface is a challenge for existing research, which is still in the initial stage.

The objective of this chapter is to master the experimental method of uniformly covering the thin film with an aimed thickness that can be applied to completely cover the nanowire-modified porous electrode. The thin film that covers the inner surface of the porous electrode uniformly can reduce the electrochemical reaction at the interface between the electrode and the aqueous solution. At the same time, it can enhance the electrode strength, slow down the electrode loss, increase the electrode life, and further lower the heavy metal releasing to the effluent.

Z. Huo, *Nanowire-assisted Flow-through Electrode Enabling Electroporation Disinfection of Reclaimed Water*, Springer Theses,
https://doi.org/10.1007/978-981-99-4502-3_5

5.2 Experimental Materials and Methods

5.2.1 Experimental Reagents

Dopamine and Tris reagents were purchased from Sigma, USA, and the rest of the reagents were the same as in Sect. 2.2.1.

5.2.2 Fabrication of Poly-Dopamine (PDA)-Coated Nanowire-Modified Electrodes

(1) CuO and Cu_3P nanowire-modified copper foam electrodes preparation.
The copper foam used in Sect. 5.2.2 was selected as the porous electrode in the nanowire-assisted electroporation disinfection device with aimed diameter (10 mm), thickness (2 mm), and pore size (200 μm). The CuO nanowire-modified copper foam electrode was prepared in the same way as in Sect. 2.2.2, and the Cu_3P nanowire-modified copper foam electrode was prepared in the same way as in Sect. 4.2.2.
(2) PDA-coated porous electrode preparation.
Dissolve 0.1 g of dopamine in 40 mL of Tris solution (0.01 M, 2.5 g/L) and adjust the solution pH to 8.5. The synthesized nanowire-modified copper foam electrode was immersed in the freshly prepared dopamine solution and reacted at 40 °C in the open air. After each 4 h reaction cycle, the electrode was removed and washed repeatedly with DI water, and then submerged again in freshly prepared dopamine solution for the next cycle. After 1, 3, 4, and 6 cycles, PDA films of different thicknesses can be uniformly covered on the electrode surface. The electrodes were washed repeatedly with DI water and dried in a drying dish.

5.2.3 Characterization of PDA-Coated Nanowire-Modified Electrodes

(1) Morphological characterization.
SEM was used to characterize the PDA-coated nanowire-modified copper foam electrode, and the characterization method was the same as in Sect. 2.2.3. TEM was used to characterize the morphology of the single PDA-coated nanowire. The sample preparation and material characterization methods were the same as in Sect. 2.2.3.
(2) Elemental and functional group analysis.
The elemental distribution analysis of the single PDA-coated nanowires was carried out by using an EDX with TEM. XPS (Kratos Analytical, K-alpha) was used to analyze the chemical composition and properties of the surface elements

of the PDA-coated nanowire-modified copper foam electrode. The experimental method is the same as in Sect. 4.2.9. Raman spectroscopy (Renishaw, inVia Reflex) was used to analyze the functional groups of the PDA-coated nanowire-modified copper foam electrode, and the measurement wavelength was chosen from 600 to 1800 cm^{-1}.

5.2.4 *Disinfection Device Construction Using PDA-Coated Nanowire-Modified Electrodes*

The construction of the disinfection device is the same as in Sect. 4.2.3.

5.2.5 *Microbes and Water Samples Used in Experiments*

Microbial strain culture and water sample configuration experimental methods are the same as in Sect. 4.2.5.

5.2.6 *Electroporation Disinfection for Microbes Using PDA-Coated Nanowire-Modified Electrodes*

(1) Disinfection efficiency analysis.
The experimental method is the same as in Sect. 2.2.5.
(2) Analysis of disinfection mechanism.
The changes in cell morphology before and after treatment were analyzed by SEM, and the experimental methods were the same as those in Sect. 2.2.9.

5.2.7 *Analysis of the Disinfection Efficiency Using Nanowire-Modified Electrodes for Long-Term Operation*

The experimental method is the same as in Sect. 4.2.8.

5.2.8 Analysis of the Loss Mechanism of Electrode During Long-Term Operation

(1) Long-term operation copper loss analysis.
The copper loss from the electrode was analyzed during long-term operation, and the concentration of copper ions release due to chemical dissolution and the concentration of copper particles lost due to water detachment were measured. The experimental method is the same as Sect. 4.2.9.
(2) Analysis of electrode morphology during long-term operation.
The change of electrode morphology during long-term operation was analyzed by SEM. The experimental method is the same as Sect. 4.2.9.
(3) Long-term operation electrode loss mechanism analysis.
To analyze the loss mechanism of the electrode during long-term operation, XRD and XPS were used to determine the change of material structure and surface elements of the electrode in different operation times. The experimental method is the same as Sect. 4.2.9.

5.3 Fabrication of PDA-Coated Nanowire-Modified Electrodes

The use of nano-scaled thin films to cover materials with nanostructures can effectively protect nanomaterials and improve chemical and physical stability. Common coating materials include metallic materials (gold, aluminum), carbon materials, conductive organic materials (polyaniline, polypyrrole, and polythiophene), and organic polymers (PDA). Among the existing surface coating materials, PDA is widely used in the field of biological and energy functional materials due to its advantages of self-polymerization in the solution system, strong bonding to the material surface, and easy handling. Therefore, in this chapter, PDA is chosen as the coating material to cover the surface of the CuO and Cu_3P nanowire-modified copper foam electrodes. The enhancement of electrode lifetime by PDA coating was also investigated.

5.3.1 Fabrication of PDA-Coated CuO Nanowire-Modified Electrodes

The experimental procedure of coating PDA films on the surface of CuO nanowire-modified copper foam electrode is shown in Fig. 5.1. The fabricated CuO nanowire-modified copper foam electrode was immersed in freshly prepared dopamine solution (2.5 g/L, 0.01 M Tris, pH = 8.5) and reacted at a constant temperature of 40 °C in

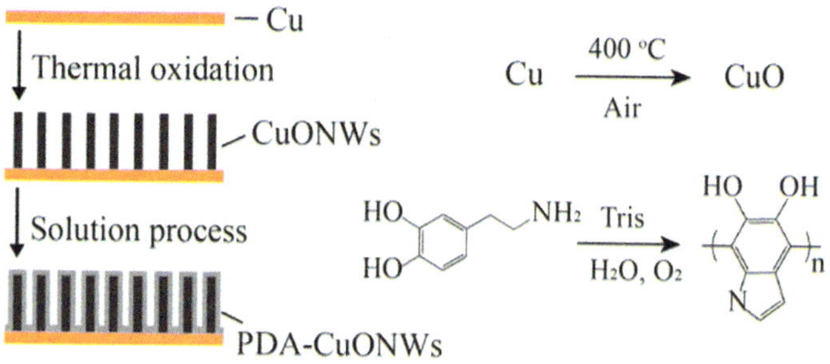

Fig. 5.1 Fabrication of PDA-coated CuO nanowire-modified electrodes

the open air for a certain time. Then, the thin PDA film with uniform thickness can be deposited on the surface of the nanowire.

The morphological characterization of PDA-coated CuO nanowire-modified copper foam electrode is shown in Fig. 5.2. The SEM image of the electrode showed that the PDA did not change the internal structure of the porous electrode during the polymerization deposition. The SEM image of the single CuO nanowires showed that the PDA coating was deposited uniformly on the surface of each nanowire. In addition, the TEM image of the single PDA-coated CuO nanowire showed that the PDA film completely covered the CuO nanowire structure. In summary, the morphological characterization of the electrode showed that the PDA uniformly and completely covered the CuO nanowire-modified copper foam electrode surface.

The process of PDA deposition is the reaction of dopamine molecules with oxygen in water solution under alkaline conditions (pH = 8.5), which continuously polymerizes into long-chain organic polymorphs. The PDA deposition thickness can be effectively controlled by controlling the reaction time. The self-polymerization reaction of dopamine molecules will not only occur on the electrode surface to form a uniform film, but it also occurs in the solution, polymerizing into small particles. These small particles will keep increasing with the reaction time and eventually deposited by gravity onto the electrode surface, affecting the flatness of the electrode surface coating. As shown in Fig. 5.3, after 12 and 24 h of continuous deposition by dopamine, a certain number of PDA particles were deposited on the nanowire surface,

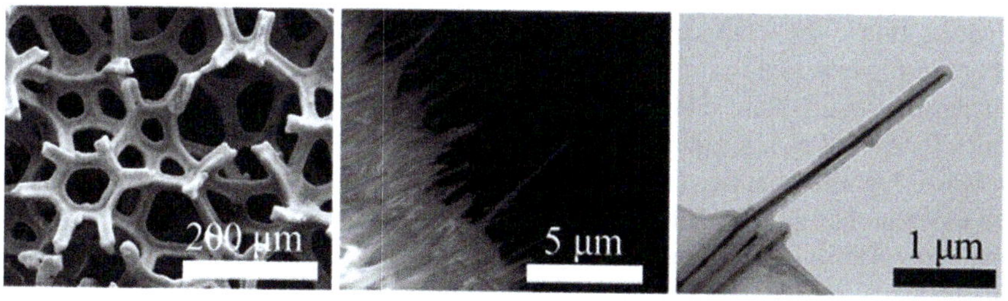

Fig. 5.2 Morphological characterization of PDA-coated CuO nanowire-modified electrodes

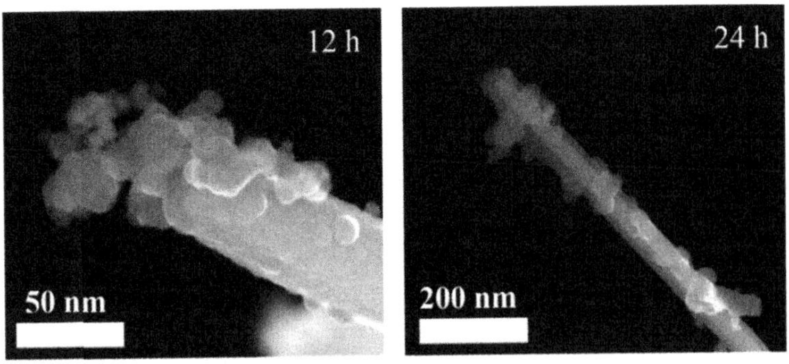

Fig. 5.3 Morphological characterization of PDA-coated CuO nanowire-modified electrodes after continuous deposition

and these particles not only seriously affected the morphology of the nanowires but also harmed the subsequent electroporation disinfection efficiency.

Therefore, to avoid the long polymerization time and the deposition of PDA particles in the solution to the electrode surface affecting the nanowire morphology, during the actual coating process, a freshly prepared dopamine solution was chosen to polymerize for 4 h as a cycle, and the PDA polymerization reaction was reconfigured at the end of each polymerization cycle (4 h) by reconfiguring the dopamine solution. Since the single reaction cycle is only 4 h, it is difficult for the dopamine molecules to form small particles by self-polymerization in the solution, and therefore the uniformity of the coating is not affected.

5.3.2 Characterization of PDA-Coated CuO Nanowire-Modified Electrodes

The PDA film deposition time can be controlled by different cycles of PDA film reaction, and the thickness of PDA coating can be controlled while ensuring the uniform and flat film. The surface morphology of PDA-coated CuO nanowires at different thicknesses after different reaction times is shown in Fig. 5.4. At different coating times, the PDA can completely wrap the entire nanowire structure. The relationship between coating time and coating thickness was found to be linear with increasing coating time, and the PDA thickness increased by 4.7 nm per hour. 110 nm thick PDA films covered the nanowire surface uniformly after 24 h (6 coating cycles). It is worth noting that two (or more) nanowires in relatively close proximity may be wrapped in the same PDA film as the coating time grows and the coating thickness increases. This bundle structure has better physical stability and is expected to exhibit a longer lifetime during the subsequent long-term disinfection.

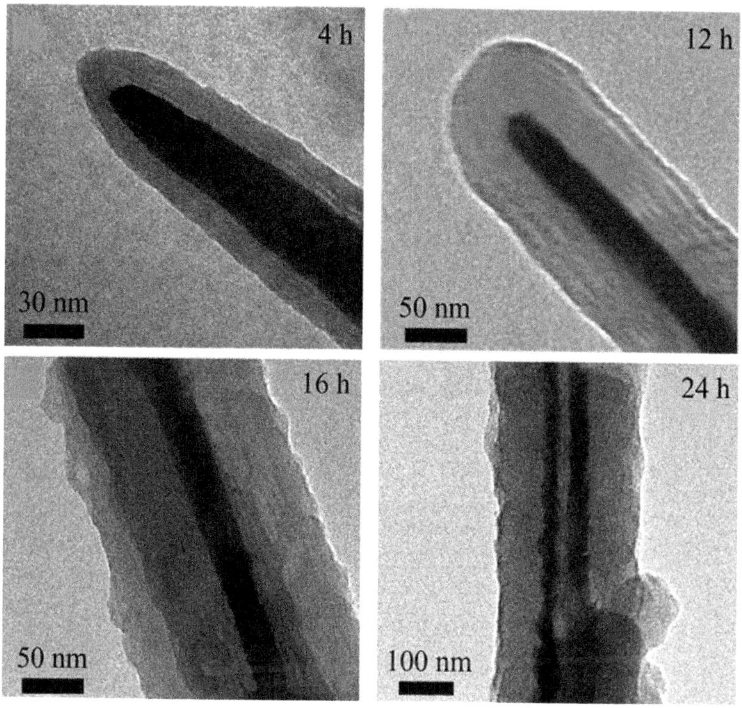

Fig. 5.4 Surface morphology of PDA-coated CuO nanowires with different coating time

Covering the nanowire surface with a nonconductive nanoscale film does not significantly affect the electric field enhancement phenomenon with the strong electric field distribution region. Since bacteria can be inactivated by entering the strong electric field region during electroporation, they do not need to contact with the nanowires. Theoretically, the PDA-coated CuO nanowire-modified copper foam electrodes can still achieve efficient electroporation disinfection treatment.

To further demonstrate the successful coating of PDA on CuO nanowires and its tightly wrapping of entire nanowire structure, the study utilized TEM with its accompanying EDX for interfacial morphology analysis and elemental surface analysis. As shown in Fig. 5.5, the elemental surface analysis of PDA-coated CuO nanowire showed that the C and O containing elements (representing PDA) are uniformly covered on the surface of the Cu containing elements (representing the CuO nanowire). In addition, the analysis of the interface between the PDA film and CuO nanowire showed that the PDA coating covered closely on the surface of CuO nanowire, which proves that the PDA coating wrapped the nanowire structure closely and has a strong bonding force.

Finally, the study used XPS to analyze the surface elements of the PDA-coated CuO nanowire-modified copper foam electrode. As shown in Fig. 5.6, the pristine CuO nanowire-modified copper foam electrode without PDA coating only has the

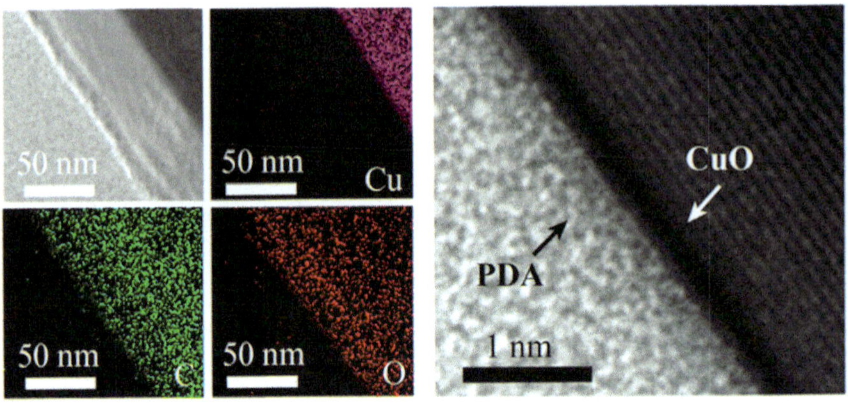

Fig. 5.5 Elemental analysis and interface characterization of PDA-coated CuO nanowires

Fig. 5.6 Elemental analysis of PDA-coated CuO nanowire-modified electrodes

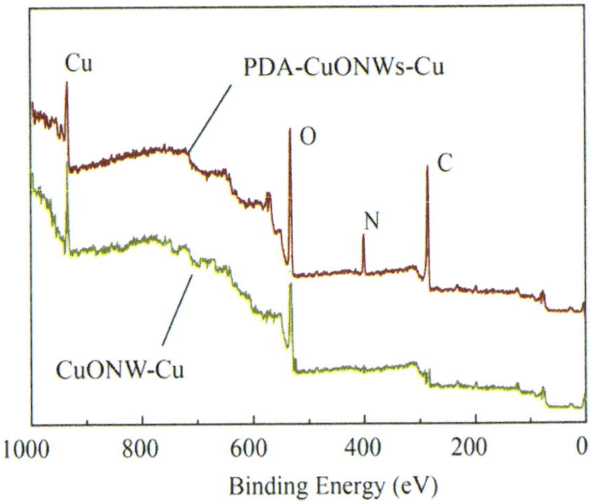

elements of Cu and O. Nevertheless, additional elemental peaks of C and N were detected for the PDA-coated CuO nanowire-modified electrode, indicating that PDA was successfully deposited onto the surface of CuO nanowire-modified copper foam electrode.

5.3.3 *Fabrication of PDA-Coated Cu_3P Nanowire-Modified Electrodes*

This study also wrapped a PDA protective film on the surface of the Cu_3P nanowire-modified copper foam electrode, and the preparation process is shown in Fig. 5.7.

Similar to the process of covering PDA protective film on CuO nanowires, the Cu_3P nanowire-modified copper foam electrode prepared by the two-step process was immersed in a freshly prepared PDA solution and reacted at a constant temperature (40 °C). The thin PDA film with uniform thickness can be deposited on the surface of the nanowire after a specific cycle of reaction.

The morphological characterization of the PDA-coated Cu_3P nanowire-modified copper foam electrode is shown in Fig. 5.8. The SEM morphology of the electrode showed that the PDA film does not affect the porous structure of copper foam, and the PDA film covered the whole surface of Cu_3P nanowires uniformly.

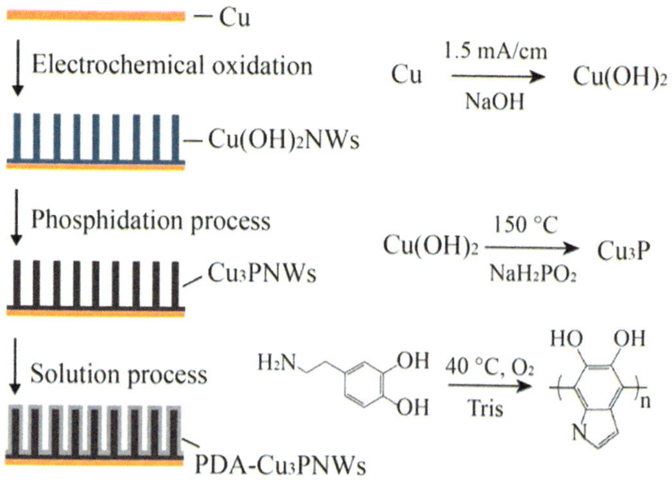

Fig. 5.7 Fabrication of PDA-coated Cu_3P nanowire-modified electrodes

Fig. 5.8 Morphological characterization of PDA-coated Cu_3P nanowire-modified electrodes

5.3.4 Characterization of PDA-Coated Cu_3P Nanowire-Modified Electrodes

As shown in Fig. 5.9, a specific cross-section of PDA-coated Cu_3P nanowires was selected for the elemental line analysis, and the C element (representing PDA film) was covered around the surface of the Cu element (representing Cu_3P nanowires). This demonstrated that PDA was uniformly wrapped around the surface of the Cu_3P nanowires.

In addition, the elemental surface analysis of the PDA-coated Cu_3P nanowire-modified electrode was performed. As shown in Fig. 5.10, where the C and O elements representing the PDA film are covered on the outer surface of the Cu and P elements (representing the Cu_3P nanowires), proving that the PDA film successfully coated on the Cu_3P nanowires. The analysis of the interface between the Cu_3P nanowire and the PDA film also showed that the PDA film closely adhered to the surface of the Cu_3P nanowires.

The relationship between the deposition time and coating thickness during PDA coating was further analyzed by TEM. As shown in Fig. 5.11, the thickness of PDA film coated on the Cu_3P nanowires increased linearly with the deposition time at a rate of 3.6 nm/h. After 24 h of deposition (6 coating cycles), the thickness of PDA film increased to 86 nm. Compared to the relatively faster deposition rate in

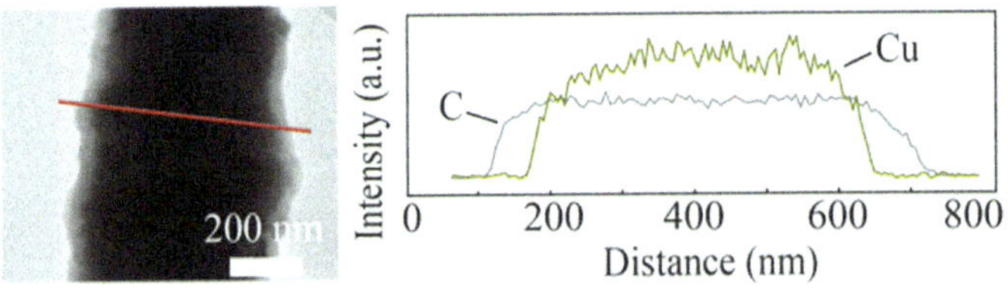

Fig. 5.9 Elemental line analysis of PDA-coated Cu_3P nanowires

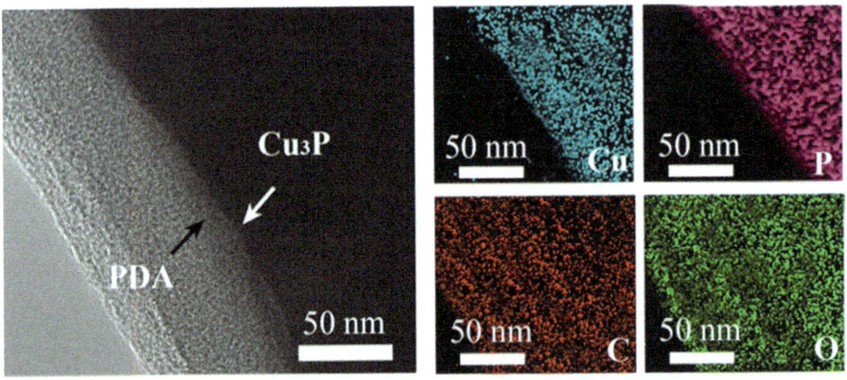

Fig. 5.10 Elemental surface and interface analysis of PDA-coated Cu_3P nanowires

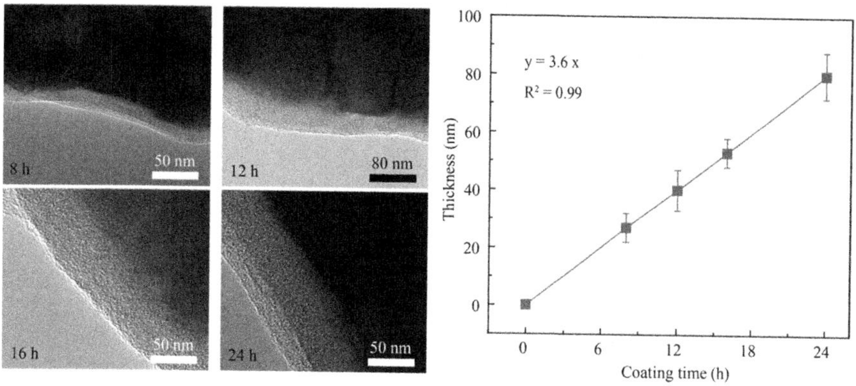

Fig. 5.11 TEM characterization showing the coating thickness of PDA film on Cu_3P nanowires

CuO nanowire coating, the deposition speed was reduced when deposited on Cu_3P nanowires due to the coarser physical structure of Cu_3P nanowires.

Surface elemental analysis of PDA-coated Cu_3P nanowire-modified copper foam electrode was performed using XPS. As shown in Fig. 5.12, the characteristic peaks of C, O, and N elements representing PDA film appeared on the surface of the electrode after coating. The elemental analysis of C showed that new C–N, C=O, and O–C=O bonds were formed after coating, corresponding to 285.5, 287.5, and 288.5 eV binding energies, respectively. According to the dopamine polymerization mechanism, the C-N is caused by the presence of amine functional groups in the PDA, while the C=O and O–C=O bonds are due to structural changes in the long-chain dopamine during the self-polymerization process [127, 153]. The analysis of the N element showed that after coating, a new peak located at 399.5 eV representing the amine functional groups appeared [154]. In addition, the functional groups on electrodes were analyzed using Raman spectroscopy analysis. In contrast to the Cu_3P nanowire-modified copper foam electrode that did not produce distinct peaks, distinct aromatic functional groups appeared at 1335 and 1568 cm^{-1} after coating, demonstrating the successful deposition of PDA on the Cu_3P nanowire-modified electrode surface [155, 156].

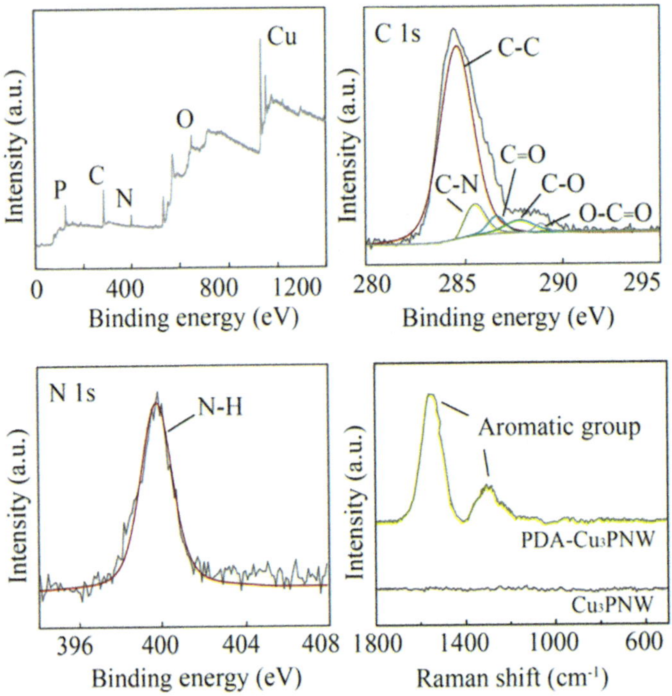

Fig. 5.12 Surface elemental analysis and functional groups investigation of PDA-coated Cu_3P nanowire-modified electrodes

5.4 Electroporation Disinfection Efficiency of PDA-Coated Nanowire-Modified Electrodes

5.4.1 Disinfection Efficiency of PDA-Coated Nanowire-Modified Electrodes

The prepared PDA-coated CuO and Cu_3P nanowire-modified copper foam electrodes were installed in an electroporation disinfection reactor to investigate the disinfection efficiency. The experiments firstly used *E. coli* as the model strain. As shown in Fig. 5.13, for the CuO nanowire-modified electrode, there was a slight decrease in disinfection efficiency with the increasing PDA thickness after PDA coating. The pristine CuO nanowire-modified copper foam electrode achieved complete inactivation of *E. coli* (log removal rate >6.0 log) at an external voltage of 1 V and HRT of 5 s. After 12 h PDA deposition, the disinfection efficiency was the same as that of the pristine CuO nanowire-modified electrode, whereas when increasing the deposition time to 16 and 24 h, the disinfection efficiency slightly reduced with the necessary HRT for complete inactivation of 6 and 8 s, respectively. It is possible that the thicker

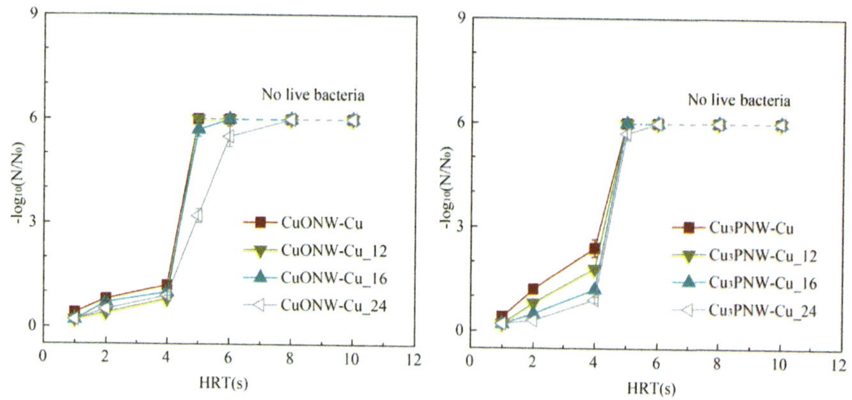

Fig. 5.13 Disinfection efficiency of PDA-coated nanowire-modified electrodes to *E. coli*

PDA film occupied part of the strong electric field area, which reduced the opportunities for bacteria entering the strong electric field and therefore eliminating the disinfection efficiency. However, the affected disinfection efficiency was still high enough for practical applications.

For Cu_3P nanowire-modified copper foam electrodes, the disinfection efficiency remained high after PDA coating, and most of them achieved complete inactivation at 1 V, 5 s. The disinfection efficiency was slightly reduced only when the deposition time was 24 h. The HRT required for complete microbial inactivation was 6 s. Compared with CuO nanowire-modified electrode, the disinfection efficiency of the Cu_3P nanowire-modified electrode was less affected by PDA coating. It is likely due to the thicker physical structure of the Cu_3P nanowire (400 nm in diameter). The PDA film with a thickness of <100 nm has a weak impact on the local electric field of Cu_3P nanowires and therefore on the disinfection performance.

The study further analyzed the disinfection efficiency of PDA-coated nanowire-modified electrodes for different bacteria and actual wastewater samples using the electrodes after 24 h PDA deposition. As shown in Fig. 5.14, for the CuO nanowire-modified electrode with 24 h PDA film deposition, complete inactivation was achieved for different model bacteria and actual wastewater samples with an external voltage of 1 V and an HRT of 10 s. The disinfection efficiency of Cu_3P nanowire-modified electrode with 24 h PDA film deposition was higher than that of PDA-coated CuO nanowire-modified electrode under the same operating conditions. The HRT required for the complete inactivation for different model bacteria and actual wastewater samples was 8 s when applying the same operating voltage (1 V).

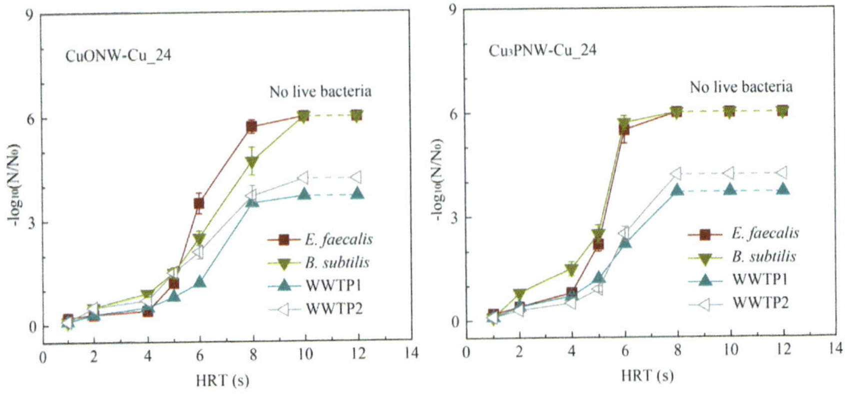

Fig. 5.14 Disinfection efficiency of PDA-coated nanowire-modified electrodes for different bacteria and actual wastewater

5.4.2 Analysis of the Disinfection Mechanism of PDA-Coated Nanowire-Modified Electrodes

To further analyze the disinfection mechanism, SEM was used to characterize the bacterial morphology before and after disinfection. As shown in Fig. 5.15, compared with the smooth and intact cell membrane structure before disinfection, the bacteria formed small pores (<200 nm in diameter) on the surface of the cell membrane after disinfection while maintaining the cell structure. The small pores with this morphology were typical electroporation pores, which proved that the disinfection mechanism of the nanowire-modified copper foam electrode did not change after PDA coating, and it was still electroporation.

5.5 Long-Term Disinfection Performance and Loss Mechanism of PDA-Coated Nanowire-Modified Electrodes

5.5.1 Long-Term Disinfection Performance of PDA-Coated Nanowire-Modified Electrodes

The long-term disinfection performance of PDA-coated CuO and Cu_3P nanowire-modified copper foam electrodes was investigated with an external voltage of 1 V and an HRT of 10 s. As shown in Fig. 5.16, for CuO nanowire-modified copper foam electrodes, the electrode lifetime for continuous disinfection was significantly improved after PDA coating. The disinfection efficiency of pristine CuO nanowire-modified electrode was significantly reduced after 15 min operation with no microbe

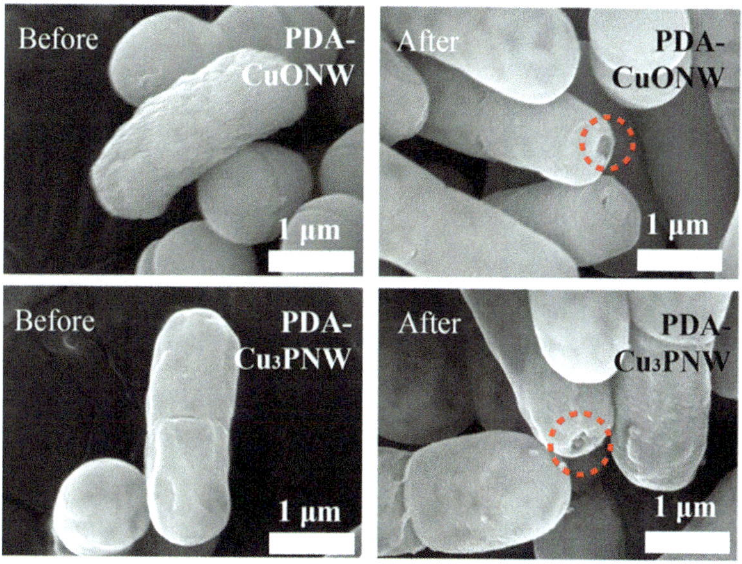

Fig. 5.15 Analysis of the disinfection mechanism of PDA-coated nanowire-modified electrodes

inactivation after operation. Nevertheless, with 16 h PDA film deposition, the CuO nanowire-modified electrode achieved complete inactivation for 4 h of continuous operation. When the PDA film deposition time increased to 24 h, the lifetime of the electroporated electrode was further improved, and complete inactivation of microbes was achieved for 12 h. When protected by PDA film, the stability of CuO nanowire-modified electrodes was significantly improved, and the electrode lifetime was increased by 36 times. For Cu_3P nanowire-modified copper foam electrode, the electrode lifetime was also significantly increased after PDA coating. The pristine Cu_3P nanowire-modified electrode enabled complete disinfection for 12 h, while the disinfection efficiency gradually decreased after 12 h. After applied 16 h PDA deposition, Cu_3P nanowire-modified electrodes achieved complete disinfection for 36 h, continuously. When the PDA film deposition time increased to 24 h, the electrode lifetime was further improved that achieved complete inactivation for 4 consecutive days (96 h). Similar to CuO nanowire-modified electrode, the stability of Cu_3P nanowire-modified electrodes protected by PDA film was significantly improved and the electrode lifetime was increased by nearly 10 times.

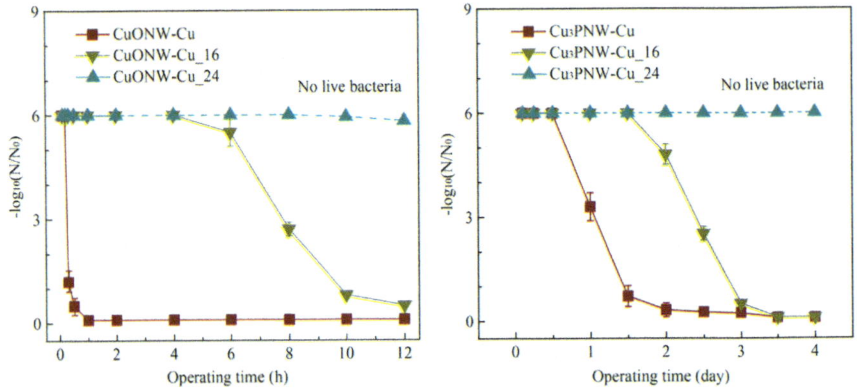

Fig. 5.16 Long-term disinfection performance of PDA-coated nanowire-modified electrodes

5.5.2 PDA-Coated Nanowire-Modified Electrode Loss Analysis

The study further analyzed the electrode loss phenomenon of the PDA-coated nanowire-modified copper foam electrode during long-term operation. The copper concentration in the effluent during the operation was measured and the results are shown in Fig. 5.17. For the CuO nanowire-modified copper foam electrode, the electrode loss was significantly reduced after PDA coating compared to the pristine electrode with severe electrode loss (340 μg/L copper concentration in the effluent). After 0.2 h of operation, the copper concentration in the effluent was 80 μg/L, which included 39 μg/L of copper particles detached by water flow and 51 μg/L of dissolved copper ions. The concentration of dissolved copper ions remained stable with time. At the same time, the concentration of copper particles detached by water flow gradually decreased, which was likely due to the absence of nanowire structure on the surface of the electrode after a long period of operation due to water scouring. For Cu_3P nanowire-modified electrodes, the electrode loss rate was also significantly reduced after PDA coating: the copper concentration in the effluent decreased from 25 to 14 μg/L after 15 min of operation under the same operating conditions compared with that without PDA coating. The loss rate gradually decreased with the increase of operation time. After 4 days of continuous operation, the same concentration in the effluent was 7.5 μg/L.

It is noteworthy that the PDA film slowed down the loss of CuO nanowire-modified electrode more than Cu_3P nanowire-modified electrode. The possible reason is that the stability of the CuO nanowire-modified electrode can be significantly improved after PDA coating due to the unstable structure, low physical strength, and poor chemical stability of CuO nanowires. For Cu_3P nanowire-modified electrodes, due to their high physical strength and chemical stability, the stability enhancement of Cu_3P nanowires by PDA film protection is not as obvious as that of CuO nanowires.

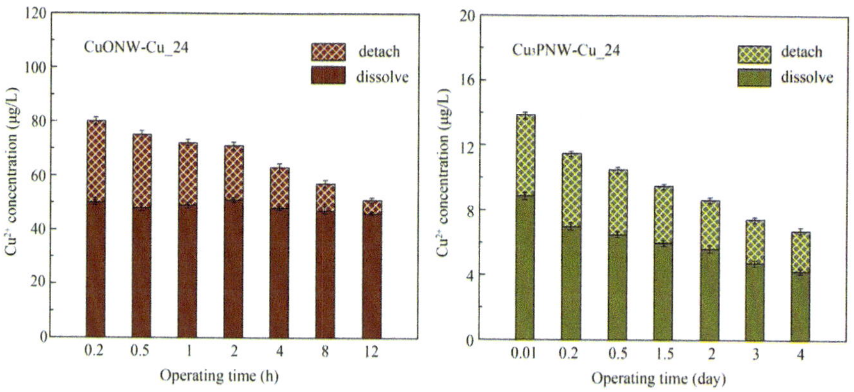

Fig. 5.17 Electrode loss phenomenon of the PDA-coated nanowire-modified copper foam electrode during long-term operation

In fact, after PDA film coating, the loss rate of Cu_3P nanowire-modified electrode was further reduced and the electrode lifetime was increased to 4 days, which has great potential to be applied in actual situations.

5.5.3 *Analysis of the Loss Mechanism of PDA-Coated Nanowire-Modified Electrodes*

(1) Analysis of electrode loss mechanism of PDA-coated CuO nanowire-modified electrode in long-term operation.

The loss mechanism of PDA-coated CuO nanowire-modified copper foam electrode was studied and analyzed during the long-term operation. The results of the morphological analysis of the electrodes during long time operation are shown in Fig. 5.18. The positive and negative electrodes showed similar morphological changes during the long-term operation. Compared with the nanowires with good morphology and high density before operation, the nanowire density decreased significantly after 0.5 h operation, and the nanowire length also decreased. As the operating time increased, the nanowire density changed less significantly, while the nanowire length decreased continuously. After 12 h of continuous operation, there was basically no nanowire structure on the electrode surface.

The loss mechanism of PDA-coated CuO nanowire-modified copper foam electrodes during the long operation time was derived from the analysis of nanowire morphology change. Since the PDA film effectively protected the CuO nanowire-modified electrode and reduced the electrochemical reaction, the main loss mechanism of the electrode was water scouring.

As shown in Fig. 5.19, during the initial operation, the physical strength of the CuO nanowires was poor, and even though coated by the PDA film, some of the nanowires

Fig. 5.18 Analysis of electrode loss mechanism of PDA-coated CuO nanowire-modified electrode during long-term operation

were still easily broken by water scouring, and the nanowire density was significantly reduced at this time (Fig. 5.20). At the same time, some of the nanowire tips were also easily broken by the water flow, resulting in the exposure of the uncoated CuO nanowires to the solution (marked by red circles in Fig. 5.21), and thus the nanowire length was also reduced (Fig. 5.20).

During the subsequent long-term operation, the remaining nanowire structure on the electrode was more stable and will not be detached by the water flow due to the protection of the PDA film (Fig. 5.20). However, the tips of CuO nanowires exposed to water underwent redox reactions and were continuously lost. Since the redox reactions occurred mainly at the tips of the nanowires exposed to water, the area of the region where the reaction occurred remained stable during the long-term operation, which kept the reaction rate consistent, leading to a consistent rate of nanowire length decreasing. In summary, PDA-coated CuO nanowires show a two-step electrode loss process. The first step is a short period in which the nanowires are detached by the water flow resulting in a significant decrease in density and a slight decrease in length, and the second step is a long-term process in which the nanowire density changes less, and the length shows a linear decrease with time (Fig. 5.20).

(2) Analysis of electrode loss mechanism of PDA-coated Cu_3P nanowire-modified electrode in long-term operation.

For the PDA-coated Cu_3P nanowire-modified electrode, the study also analyzed the electrode loss mechanism during the long-term operation. As shown in Fig. 5.22, after 2 days of continuous operation, unlike CuO nanowires, the PDA-coated Cu_3P nanowire-modified electrode still maintained the whole nanowire structure with the

Fig. 5.19 Schematic showing the electrode loss mechanism of PDA-coated CuO nanowire-modified electrode during long-term operation

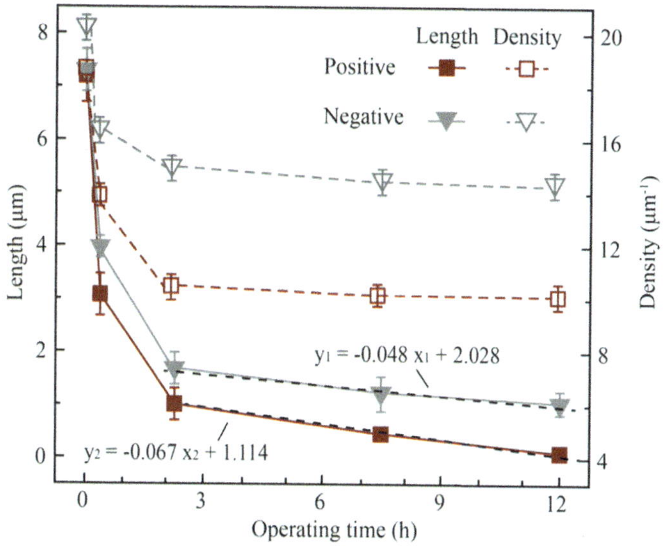

Fig. 5.20 Quantitative analysis of PDA-coated CuO nanowire-modified electrode during long-term operation

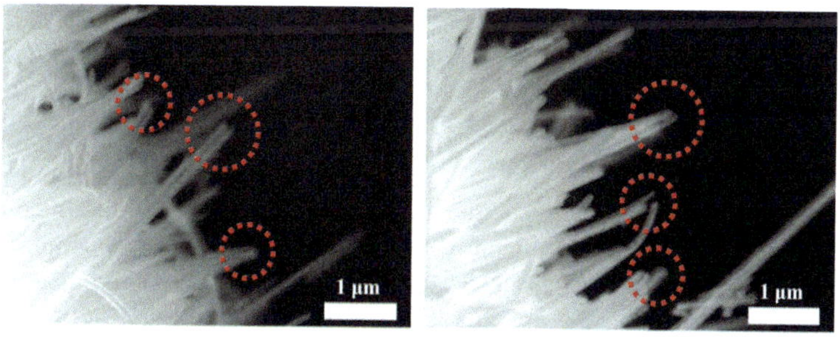

Fig. 5.21 Morphological analysis of PDA-coated CuO nanowire-modified electrode during long-term operation

less obvious density change. However, significant differences in the morphology of the positive and negative electrodes were observed during long-term operation. For the positive electrode, the nanowire tip structure was looser, probably due to the in-situ oxidation reaction at the positive electrode leading to the copper ion release. In contrast, at the negative electrode, the nanowire tips produced significant crystallization. After 4 days of continuous operation, the nanowire structure at the positive electrode decreased gradually, while the crystallization at the tip of the nanowire at the negative electrode became more obvious. These nanowires failed to maintain the high aspect ratio, which led to insufficient disinfection performance after 4 days of continuous operation.

Fig. 5.22 Morphological analysis of PDA-coated Cu_3P nanowire-modified electrode during long-term operation

The electrode loss mechanism of the PDA-coated Cu_3P nanowire-modified electrode during long operation is similar to that of the pristine Cu_3P nanowire-modified electrode. The electrochemical reaction was suppressed due to the surface PDA wrapping, and the physical strength of the electrode was further enhanced so that the electrode lifetime was increased from 12 h to 4 days. During the long-term operation, the electrochemical reaction was still the main cause of electrode loss. On the other hand, for the PDA-coated CuO nanowire-modified electrode, even after the PDA coating, the physical strength of the electrode was still not sufficient to stop the water scouring, so only 12 h of electrode lifetime was achieved. In the long-term operation, the electrode breakage by water scouring was the main reason for electrode loss, while the electrochemical reaction contributed less to the electrode loss.

5.6 Summary

(1) The physical strength and chemical stability of the nanowire-modified copper foam electrode were significantly improved by uniformly depositing a PDA film on the surface of the electrode using the solution method. The lifetime of both CuO and Cu_3P nanowire-modified copper foam electrodes was significantly increased after the PDA coating. Among them, the PDA-coated Cu_3P nanowire-modified electrode can operate continuously for 4 days and maintain the complete disinfection in water.

(2) The loss rate of electrode was significantly reduced after PDA coating, and the copper ion concentration in the effluent of PDA-coated Cu_3P nanowire-modified electrode was less than 15 ug/L.

(3) The electrode morphology analysis, electrode material structure analysis, and electrode surface element analysis revealed the electrode loss mechanism of PDA-coated nanowire-modified electrode during the long-term operation. The main reason for the loss of the CuO nanowire-modified electrode was water scouring, while the main reason for the loss of the Cu_3P nanowire-modified electrode was electrochemical corrosion. In the subsequent research using the more stable Cu_3P nanowire-modified electrode, the electrochemical reaction on the electrode surface needs to be further reduced to enhance the electrode lifetime.

Chapter 6
Altering Current Driven Nanowire-assisted Electroporation Disinfection with the Enhanced Electrode Life

6.1 Research Background

The nanowire-modified electrode can be completely covered with a tightly bonded PDA film, which significantly improves the physical and chemical stability, thus increasing the electrode lifetime. However, in the analysis of the electrode loss mechanism, even if the electrode is completely coated by the PDA film, the nanowire-modified electrode still suffers from a weak electrochemical reaction during long-term operation. As the operation time increases, the electrochemical reaction will accumulate and eventually become the main mechanism of electrode loss. Electrode surface coating technology can hardly eliminate the electrochemical reactions completely. Considering that the electrochemical reaction has a specific response time (>1 ms) to the externally applied voltage and is different from the corresponding time for electroporation disinfection (>1 μs). Therefore, theoretically, the application of altering current (AC) power at the appropriate frequency can fundamentally prevent the electrochemical reaction while achieving efficient electroporation disinfection, slowing down electrode loss and further improving electrode lifetime. However, the above-mentioned high-frequency AC power supply mode has not been reported for water treatment applications, and related research is still in its infancy. The objective of this study is to construct a nanowire-assisted electroporation disinfection device using the high-frequency AC power supply to achieve efficient electroporation disinfection while preventing the electrochemical reactions and enhancing the electrode lifetime.

Z. Huo, *Nanowire-assisted Flow-through Electrode Enabling Electroporation Disinfection of Reclaimed Water*, Springer Theses,
https://doi.org/10.1007/978-981-99-4502-3_6

6.2 Experimental Materials and Methods

6.2.1 Experimental Reagents

The reagents were the same as in Sect. 5.2.1.

6.2.2 Fabrication of PDA-Coated Nanowire-Modified Electrodes

The electrode fabrication process was the same as in Sect. 5.2.2.

6.2.3 Disinfection Device Construction Using PDA-Coated Nanowire-Modified Electrodes

A waveform generator (RIGOL, DG5000) was used as the AC power supply, and the construction of the disinfection device is the same as in Sect. 5.2.4.

6.2.4 Microbes and Water Samples Used in Experiments

Microbial strain culture and water sample configuration experimental methods are the same as in Sect. 4.2.5.

6.2.5 Electroporation Disinfection for Microbes Using PDA-Coated Nanowire-Modified Electrodes

A waveform generator (RIGOL, DG5000) was used as the AC power supply. External voltages with different frequencies were applied to the device with the fixed peak-to-peak voltage of 1 V. The rest method of the device construction and operation is the same as in Sect. 2.2.5.

6.2.6 Analysis of the Disinfection Efficiency Using Nanowire-Modified Electrodes for Long-Term Operation

External voltages generated by AC power supply (RIGOL, DG5000) with different frequencies (10 to 3.5×10^8 Hz) were applied to the device with the fixed peak-to-peak voltage of 1 V. The peak current in the circuit was measured during operation, and the rest experimental method is the same as in Sect. 4.2.8.

6.2.7 Analysis of the Loss Mechanism of Electrode During Long-Term Operation

(1) Long-term operation copper loss analysis.

The copper loss from the electrode was analyzed during long-term operation, and the concentration of copper ions release due to chemical dissolution and the concentration of copper particles lost due to water detachment were measured. The experimental method is the same as Sect. 4.2.9.

(2) Analysis of electrode morphology during long-term operation.

The change of electrode morphology during long-term operation was analyzed by SEM. The experimental method is the same as Sect. 4.2.9.

6.3 Analysis of the Disinfection Efficiency of Nanowire-assisted Electroporation Driven by a High-Frequency AC Power Supply

6.3.1 Disinfection Performance of the High-Frequency AC Driven Nanowire-assisted Electroporation

The PDA-coated CuO and Cu_3P nanowire-modified copper foam electrodes were installed in the electroporation device, and the AC power was applied to the two electrodes to determine the effect of different AC frequencies on the electroporation disinfection efficiency. As shown in Fig. 6.1, the PDA-coated CuO nanowire-modified electrode achieved complete disinfection with AC frequency up to 10^6 Hz. When the AC frequency was higher than 10^3 Hz, the current in the circuit decreased significantly, indicating a reduction of the electrochemical reaction. Further increasing the AC frequency to 10^6 Hz, the current in the circuit was further reduced to 1.5 μA, indicating that almost no electrochemical reaction occurred

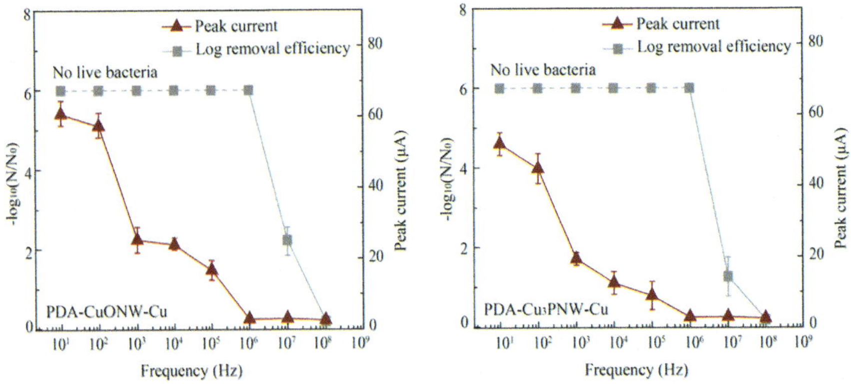

Fig. 6.1 Disinfection performance of the high-frequency AC driven nanowire-assisted electroporation

in the whole system. Therefore, at an AC frequency of 10^6 Hz, the PDA-coated CuO nanowire-modified electrode guaranteed efficient disinfection while eliminated electrochemical reactions.

For the PDA-coated Cu_3P nanowire-modified electrode, the disinfection efficiency and the variation of the current in the circuit with AC were similar to those of the PDA-coated CuO nanowire-modified electrode. When external output with frequency was less than 10^6 Hz, the PDA-coated Cu_3P nanowire-modified electrode achieved complete disinfection. The current in the circuit was extremely low (0.8 μA), indicating that there was basically no electrochemical reaction in the reaction system. In summary, for the PDA-coated nanowire-modified copper foam electrode, a high-frequency AC at 10^6 Hz can be used to drive the treatment process to ensure the disinfection efficiency while inhibiting the occurrence of electrochemical reactions, further enhancing the electrode lifetime.

Efficient disinfection can be achieved at high-frequency AC (10^6 Hz), demonstrating the low contribution of electrochemical processes to disinfection during operation. The efficiency of bacterial disinfection was significantly reduced by further increasing the AC frequency. At this time, electroporation was difficult to achieve at AC frequencies higher than 10^6 Hz, as such the disinfection efficiency decreased significantly after eliminating the effect of electroporation. This further demonstrated the microbial disinfection mechanism as electroporation. The study also used SEM to analyze the changes in microbial morphology with 10^6 Hz AC treatment. As shown in Fig. 6.2, compared to the *E. coli* with intact and smooth cell membranes before operation, the cell surface formed obvious perforations after 10^6 Hz AC treatment. These small holes with diameters < 200 nm were typical electroporation pores.

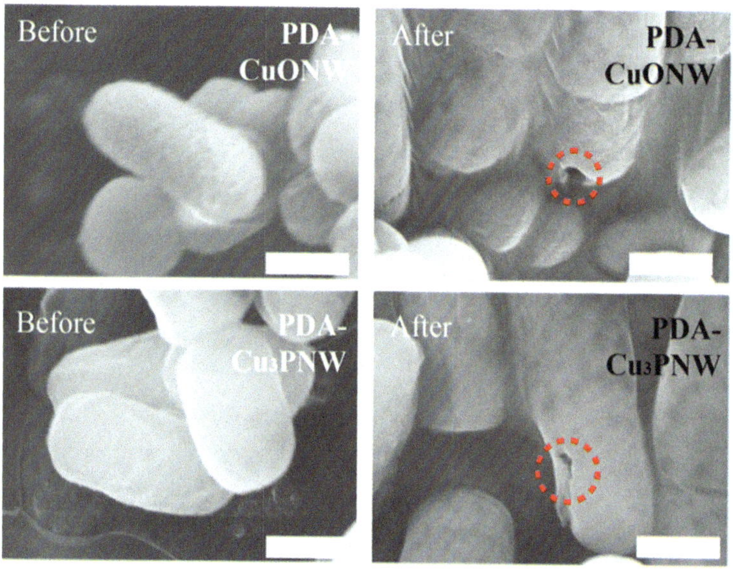

Fig. 6.2 Analysis of the disinfection mechanism of with high-frequency AC power supply

6.4 Long-Term Disinfection Efficiency and Loss Mechanism of Nanowire-assisted Electroporation Powered by High-Frequency AC

6.4.1 High-Frequency AC-Powered Nanowire-assisted Electroporation for Long-Term Disinfection

The prepared PDA-coated nanowire-modified electrode was loaded in an electroporation device and a high-frequency AC power supply with a peak current of 1 V at a frequency of 10^6 Hz was connected to electrodes. The long-term disinfection efficiency of the nanowire-assisted electroporation was investigated and as shown in Fig. 6.3, the high-frequency AC significantly improved the electrode lifetime. The high-frequency AC-powered PDA-coated CuO nanowire-modified electrode achieved complete disinfection for 2 days. While, under the same operating conditions, the DC-powered PDA-coated CuO nanowire-modified electrode only achieved complete disinfection for 12 h. The electrochemical reaction at the electrode during the disinfection process was effectively inhibited by using appropriate high-frequency AC, which increases the lifetime of the original electrode by 4 times. For the PDA-coated Cu_3P nanowire-modified electrode, complete disinfection was achieved within 15 days of continuous operation using an AC power supply. Under the same operating conditions, the DC-powered PDA-coated Cu_3P nanowire-modified

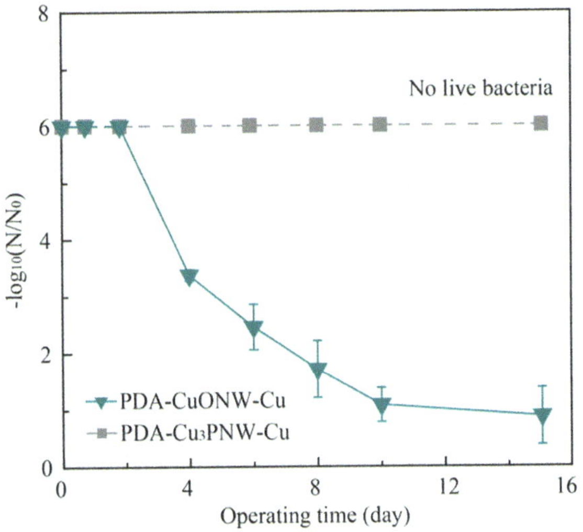

Fig. 6.3 High-frequency AC-powered nanowire-assisted electroporation for long-term disinfection

electrode only achieved complete disinfection for 4 days. These results further indicate that the high-frequency AC enhanced the electrode lifetime by reducing the electrochemical reaction. In addition, coupled with the good physical stability, the PDA-coated Cu_3P nanowire-modified electrode gained up to 15 days electrode lifetime, making the nanowire-assisted electroporation disinfection with great potential for actual situations.

6.4.2 Analysis of Electrode Loss in Long-Term Operation When Powered by High-Frequency AC

The study further determined the loss of PDA-coated nanowire-modified electrodes applied with high-frequency AC (10^6 Hz). As shown in Fig. 6.4, for both PDA-coated CuO and Cu_3P nanowire-modified electrodes, no dissolved copper ions caused by electrochemical reaction were detected in the effluent. This result further proved that the high-frequency AC power supply effectively prevented the electrochemical reaction. Meanwhile, the concentration of copper particles detached by water flow also decreased with the increasing operation time.

For the PDA-coated CuO nanowire-modified electrode, the concentration of copper particles detached by water scouring decreased from 13.2 μg/L at the initial 30-min (0.02 days) operation to 6.1 μg/L after 15 days of operation. For the PDA-coated Cu_3P nanowire-modified electrode, the electrode loss phenomenon was similar to that of the PDA-coated CuO nanowire-modified electrode, which decreased with the increase of operation time. However, the loss rate of the detached copper particles from the PDA-coated Cu_3P nanowire-modified electrode was significantly lower than that of the PDA-coated CuO nanowire-modified electrode. During

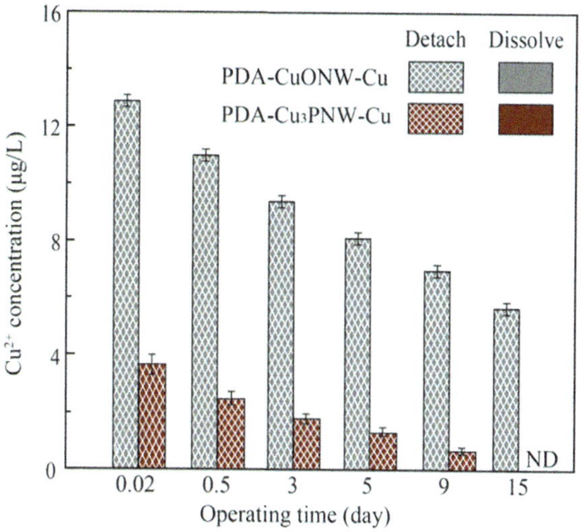

Fig. 6.4 Analysis of electrode loss in long-term operation when powered by high-frequency AC

the initial 30 min of operation, the detached copper particle on the PDA-coated Cu_3P nanowire-modified electrode was only 3.7 μg/L, which was 28% of that on the PDA-coated CuO nanowire-modified electrode under the same operating conditions. With the increase of operation time, the copper particle concentration in the effluent decreased continuously, and after 15 days of continuous operation, the detached copper particle in the effluent was below the detection limit (<0.1 μg/L). Due to the stronger physical structure of the PDA-coated Cu_3P nanowire, the loss rate of the electrode slowed down during the long operation, ensuring a longer electrode lifetime.

The morphological analysis of the PDA-coated Cu_3P nanowire-modified electrode during long-term operation with a high-frequency AC power supply was also studied. As shown in Fig. 6.5, the morphology of the PDA-coated Cu_3P nanowire did not change significantly when applied with high-frequency AC due to the effective prevention of the electrochemical reaction. After 2 days of continuous operation, the good nanowire morphology was maintained. After upgrading the operation time to 15 days, the nanowires still had good morphology with a long length (10 μm), and only some of the nanowires showed bending phenomenon.

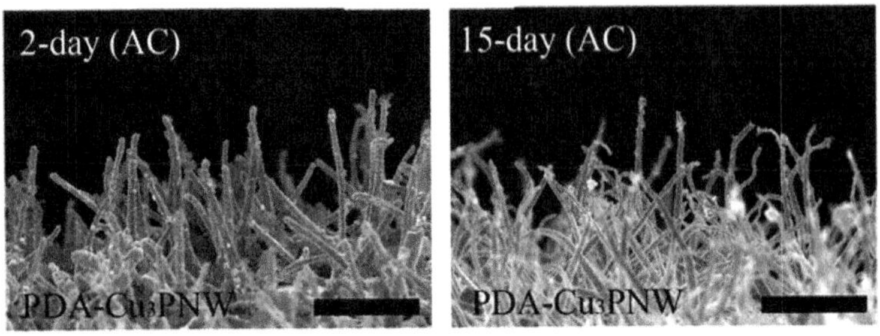

Fig. 6.5 Morphological analysis of the PDA-coated nanowire-modified electrode during long-term operation with a high-frequency AC

6.5 Summary

(1) The use of high-frequency AC (10^6 Hz) achieved efficient electroporation disinfection while preventing electrochemical reactions. It effectively eliminated electrode loss by eliminating the electrochemistry and thus, enhanced electrode lifetime. The PDA-coated Cu_3P nanowire-modified copper foam electrode achieved complete disinfection for 15 consecutive days when powered by high-frequency AC (10^6 Hz).

(2) Electrochemical reactions were effectively prevented by using a high-frequency AC power supply. No copper ions lost were detected in the effluent due to the inhibition of electrochemical reactions, and only copper particles detached by water scouring were detected. Due to the good physical strength of PDA-coated Cu_3P nanowires, the concentration of copper released in water was less than 3.7 ug/L. At this point, the electrode loss was trace and the treated water was safe.

(3) Electrode morphology analysis revealed that the high-frequency AC power supply effectively prevented the occurrence of electrochemical reactions during long-term operation.

Chapter 7
Nanowire-assisted Electroporation Disinfection for Reclaimed Water

7.1 Research Background

The aforementioned study used a high-frequency AC power supply to drive efficient microbial disinfection using PDA-coated nanowire-modified electrodes for 15 consecutive days. The long electrode lifetime of 15 days shows a strong potential for practical applications. However, existing studies have been conducted in DI water, whereas the study on nanowire-assisted electroporation disinfection technology for the real reclaimed water is still limited. Considering the complex and fluctuating quality of reclaimed water, which contains a certain concentration of organic matter and various inorganic ions, a systematic consideration of the various components of reclaimed water is necessary. Therefore, it is essential to systematically investigate the effect of various water quality indicators on the efficiency of nanowire-assisted electroporation disinfection in reclaimed water. The objective of this chapter is to clarify the effect of reclaimed water quality on nanowire-assisted electroporation disinfection. The research includes the evaluation of nanowire-assisted electroporation disinfection efficiency under actual reclaimed water conditions and the investigation of the applicability of nanowire-assisted electroporation disinfection under reclaimed water quality conditions.

Z. Huo, *Nanowire-assisted Flow-through Electrode Enabling Electroporation Disinfection of Reclaimed Water*, Springer Theses,
https://doi.org/10.1007/978-981-99-4502-3_7

7.2 Experimental Materials and Methods

7.2.1 Experimental Reagents

Alanine, arginine, glucose, soluble starch, peptone, alginate, and bovine serum albumin (BSA) were purchased from Sigma, USA, humic acid was purchased from International humic substances society, and the rest of the reagents were the same as in Sect. 2.2.1.

7.2.2 Fabrication of Nanowire-Modified Electrodes

CuO nanowire-modified copper foam electrode was chosen as the experimental electrode for the study, and the electrode preparation method was the same as in Sect. 5.2.2.

7.2.3 Disinfection Device Construction Using Nanowire-Modified Electrodes

DC-powered system: device preparation method was the same as in Sect. 5.2.4.

AC-powered system: device preparation method was the same as in Sect. 6.3.2.

7.2.4 Microbes and Water Samples Used in Experiments

(1) Water samples with different pH.
The pH of the water samples was adjusted (3 to 11) using 1 M HCl and 1 M NaOH, and the final solution ionic strength was adjusted to 10 mM using NaCl.
(2) Water samples with different ionic strengths.
Dissolve different masses of NaCl in DI water and adjust the ionic strength of the solution to 1, 5, 10, 50, and 100 mM.
(3) Water samples containing different anions and cations.
Prepare NaCl, KCl, $CaCl_2$, and Na_2SO_4 solutions with ionic strength of 10 mM.
(4) Water samples containing different types of organic substances.
Dissolve certain concentrations of alanine, arginine, glucose, soluble starch, peptone, humic acid, alginate, and BSA in DI water. Prepare the organic matter solutions with TOC of 1, 5, and 10 mg/L, respectively, and maintain the solution ionic strength of 10 mM using NaCl.
E. coli was cultured according to the strain culture method in Sect. 2.2.4 and diluted into each water sample for subsequent disinfection experiments.

Table 7.1 Basic information of wastewater reclamation treatment plant

Name	Biological treatment process
XJH	A^2O
BXH	MBR
YZ	SBR
QH	A^2O

(5) Actual reclaimed water samples.
Secondary effluent samples from municipal wastewater plant: the experimental water samples were collected from the secondary effluent (without disinfection) of four different municipal wastewater reclamation treatment plants in Beijing, which were selected attributed to their large treatment scale (>10,000 t/day) and continuous operation to ensure the representativeness of the collected water samples. The sampling method is the same as Sect. 2.2.4. The information of each wastewater reclamation treatment plant is shown in Table 7.1.

The bacteria were cultured according to Sect. 4.2.5, and the cultured *E. coli* were diluted into the secondary effluent after membrane flirtation with the final concentrations of 10^9, 10^7, 10^5, and 10^3 CFU/mL. The disinfection efficiency of nanowire-assisted electroporation disinfection technology was determined for different concentrations of *E. coli* in the secondary effluent. The disinfection efficiency of nanowire-assisted electroporation disinfection was also determined using different pure bacteria diluted in secondary effluent after membrane flirtation.

Determine the disinfection efficiency of nanowire-assisted electroporation disinfection on the actual water bodies from the natural environment. Actual water samples were selected from typical urban water bodies, taken from the Olympic Forest Park water system and Gaobeidian Lake, respectively. Some of the supplemental water sources for the Olympic Forest Park water system and Gaobeidian Lake came from the municipal wastewater reclamation treatment plant. Due to the low concentration of microorganisms in the actual water samples, the cultured bacterial suspensions were diluted in the water samples before the disinfection experiments to determine the disinfection efficiency under the actual water quality conditions. In addition, the cultured *E. coli* suspensions were diluted in the water samples with the final concentrations of 10^9, 10^7, 10^5, and 10^3 CFU/mL. The disinfection efficiency of nanowire-assisted electroporation disinfection on *E. coli* under actual water conditions was also determined.

7.2.5 Electroporation Disinfection for Reclaimed Water

DC-powered system: make water samples pass through the electroporation reactor with specific HRT, the experimental method is the same as Sect. 2.2.5.

AC-powered system: the power source is a waveform generator (RIGOL, DG5000), using designed AC output with a peak voltage of 1 V and a frequency of 10^6 Hz. Make different water samples pass through the electroporation reactor with a specific HRT. The rest of the experimental methods with Sect. 2.2.5.

7.3 Effect of Typical Reclaimed Water Quality on Nanowire-assisted Electroporation Disinfection Efficiency

7.3.1 *Effect of Inorganic Water Parameters on the Efficiency of DC-Powered Nanowire-assisted Electroporation Disinfection*

In this study, CuO nanowire-modified copper foam electrodes were used as experimental electrodes to investigate the effect of typical reclaimed water quality on the efficiency of DC and AC-powered nanowire-assisted electroporation disinfection. The effect of inorganic water parameters in the reclaimed water on DC-powered nanowire-assisted electroporation disinfection was shown in Fig. 7.1.

(1) Effect of pH.
Nanowire-assisted electroporation disinfection showed good performance in a broad range of pH, and pH fluctuations have a weak effect on the disinfection efficiency. When the HRT was fixed at 3 s, the microbial inactivation by nanowire-assisted electroporation was not significant due to the fast water flow rate. When the HRT was increased to 5 s, nanowire-assisted electroporation disinfection achieved complete disinfection over a wide pH range (log removal rate > 6.0). The efficiency of nanowire-assisted electroporation disinfection was slightly lower (~5.5 log) only when pH = 5. When the HRT was further increased to 10 s, all microorganisms in the water were completely inactivated.

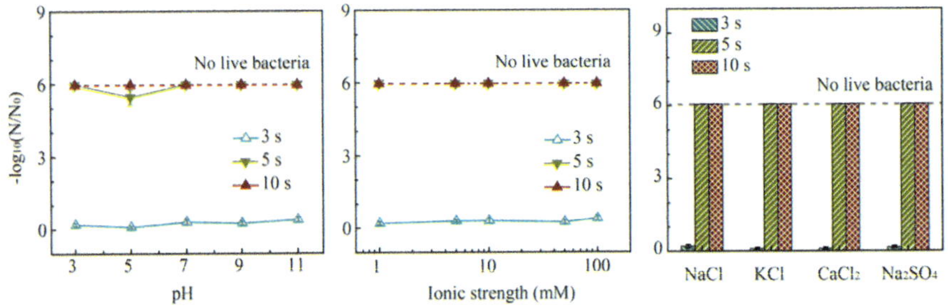

Fig. 7.1 Effect of inorganic water parameters on the efficiency of DC-powered nanowire-assisted electroporation disinfection

(2) Effect of ionic strength.
The effect of ionic strength on the disinfection efficiency of nanowire-assisted electroporation was also not significant. When the HRT was 3 s, the disinfection efficiency nanowire-assisted electroporation was low due to the fast flow rate. When the HRT was increased to 5 s, nanowire-assisted electroporation disinfection achieved complete disinfection (log removal rate > 6.0) within a wide range of ionic strengths. When the HRT was further increased to 10 s, all the microbes were inactivated.

(3) Effect of ion type.
The effect of different ion types on nanowire-assisted electroporation disinfection was also small. When the HRT was 3 s, the disinfection efficiency was not significant due to the fast flow rate. When the HRT was increased to 5 s, nanowire-assisted electroporation disinfection achieved complete disinfection in NaCl, KCl, $CaCl_2$, and Na_2SO_4 solutions, respectively (log removal rate > 6.0). The anion and cation types have less impact on nanowire-assisted electroporation disinfection. When the HRT was further increased to 10 s, the microbes were completely inactivated in different solutions.

7.3.2 *Effect of Inorganic Water Parameters on the Efficiency of AC-Powered Nanowire-assisted Electroporation Disinfection*

The study further analyzed the effect of inorganic factors in the reclaimed water on the efficiency of AC-powered nanowire-assisted electroporation disinfection. As shown in Fig. 7.2, similar to the DC-powered system, changes in pH, ionic strength, and ion type showed no significant effect on the AC-powered nanowire-assisted electroporation disinfection performance. The AC-powered nanowire-assisted electroporation achieved complete disinfection in solutions containing different pH, ionic strength, and ion type at an HRT of 5 s (log removal rate > 6.0).

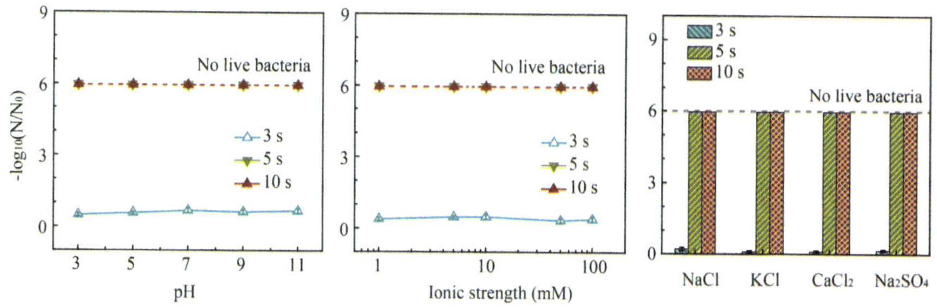

Fig. 7.2 Effect of inorganic water parameters on the efficiency of AC-powered nanowire-assisted electroporation disinfection

In summary, the effect of inorganic factors on nanowire-assisted electroporation disinfection in reclaimed water was relatively limited, and both DC and AC powered systems achieved efficient microbial inactivation over a wide range of inorganic factors. The reason why the nanowire-assisted electroporation disinfection was less affected by inorganic water quality factors was due to the specific physical-based electroporation mechanism. Inorganic water quality factors have less influence on the localized electric field distribution at the tip of the nanowire, and therefore had less influence on the electroporation disinfection efficiency.

7.3.3 Effect of Organic Matter on the Disinfection Efficiency of DC-Powered Nanowire-assisted Electroporation

Typical organic compounds, including alanine, arginine, glucose, soluble starch, peptone, humic acid, alginate, and BSA, were selected to represent the different classes of organic compounds that may exist in reclaimed water. Among them, alanine, arginine, and glucose represent small molecules (amino acids and sugars), soluble starch and peptone represent medium molecular weight organic compounds (polysaccharides and peptides), humic acid represents the common soluble organic compounds with large molecular weight in water while alginate and BSA represent large molecular weight organic compounds (polysaccharides and proteins). The effect of different categories and concentrations of organic matter on nanowire-assisted electroporation disinfection efficiency was analyzed. As shown in Fig. 7.3, for small molecules such as alanine, arginine, and glucose, there was no effect on the efficiency of DC-powered nanowire-assisted electroporation at different concentrations (1–10 TOC mg/L). For the higher molecular mass organics, the overall disinfection efficiency decreased with the concentration of the increasing organic. For soluble starch, peptone, and humic acid with medium molecular mass, only at higher concentrations (10 TOC mg/L) had some impact on disinfection efficiency, and the disinfection performance was able to further improve by appropriately extending the HRT. For alginate and BSA, which have larger molecular weights, there was a significant inhibition on nanowire-assisted electroporation disinfection at higher concentrations. At a concentration of 10 TOC mg/L, nanowire-assisted electroporation disinfection only achieved 1.2 log removal of microbes even when the HRT was extended to 10 s.

Therefore, high concentrations of macromolecular organics may affect nanowire-assisted electroporation disinfection efficiency. The study further analyzed the effect of organic matter properties on electroporation disinfection. As shown in Fig. 7.4, at high concentrations (10 TOC mg/L), small molecular weight organics (alanine, arginine, and glucose) had no effect on nanowire-assisted electroporation disinfection efficiency, while all these small molecular organics had high solubility ($>10^5$ mg/L). For the organic matters with higher molecular weight (soluble starch, peptone, humic acid, alginate, and BSA), there was an impact on the electroporation disinfection

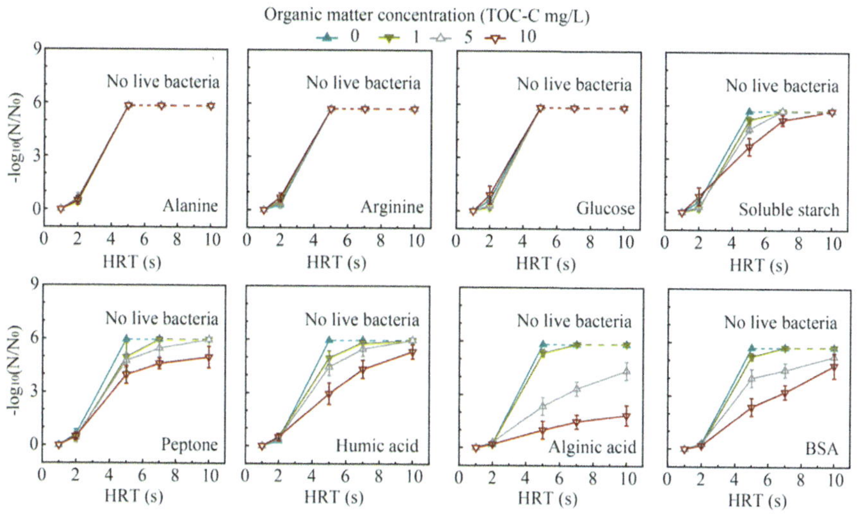

Fig. 7.3 Effect of organic matter on the disinfection efficiency of DC-powered nanowire-assisted electroporation

when the high concentration was high. These organic matters with large molecular weights were able to divide into two categories according to the degree of their influence on disinfection efficiency. The first group was the organic matter with medium molecular weight (<10^4 g/mol), including soluble starch, peptone, and humic acid. Nanowire-assisted electroporation achieved > 3.0 log microbial removal in the presence of these substances. Although the above three organics had the same molecular weight, they showed a slightly different effect on the disinfection efficiency. Soluble starch had the lowest impact (log removal rate 4.0 log), followed by peptone (log removal rate 3.7 log), and humic acid showed the severest impact on the disinfection efficiency (log removal rate 3.0 log). The effect of these organic matters on the disinfection efficiency was consistent with the organic solubility: soluble starch having the largest solubility among the three substances (7×10^5 mg/L), followed by peptone (4×10^4 mg/L), and humic acid having the lowest solubility (2×10^3 mg/L).

The second group of organic matters with higher molecular weight (>10^4 g/mol), including alginate and BSA, had a more significant impact on the disinfection efficiency. For these two substances, the impact on the disinfection efficiency of nanowire-assisted electroporation was similar to that of the medium molecular weight organics (soluble starch, peptone, and humic acid). Under similar molecular weight, the impact of lower solubility on disinfection efficiency was more significant. The solubility of alginate was only (10^3 mg/L), which was lower than that of BSA (3×10^4 mg/L), and the disinfection efficiency was only 1.0 log in the high concentration of alginate solution, which was lower than that of BSA solution (2.0 log) under the same conditions.

In summary, the molecular weight of organic matter is the main factor affecting nanowire-assisted electroporation disinfection, and the disinfection efficiency is

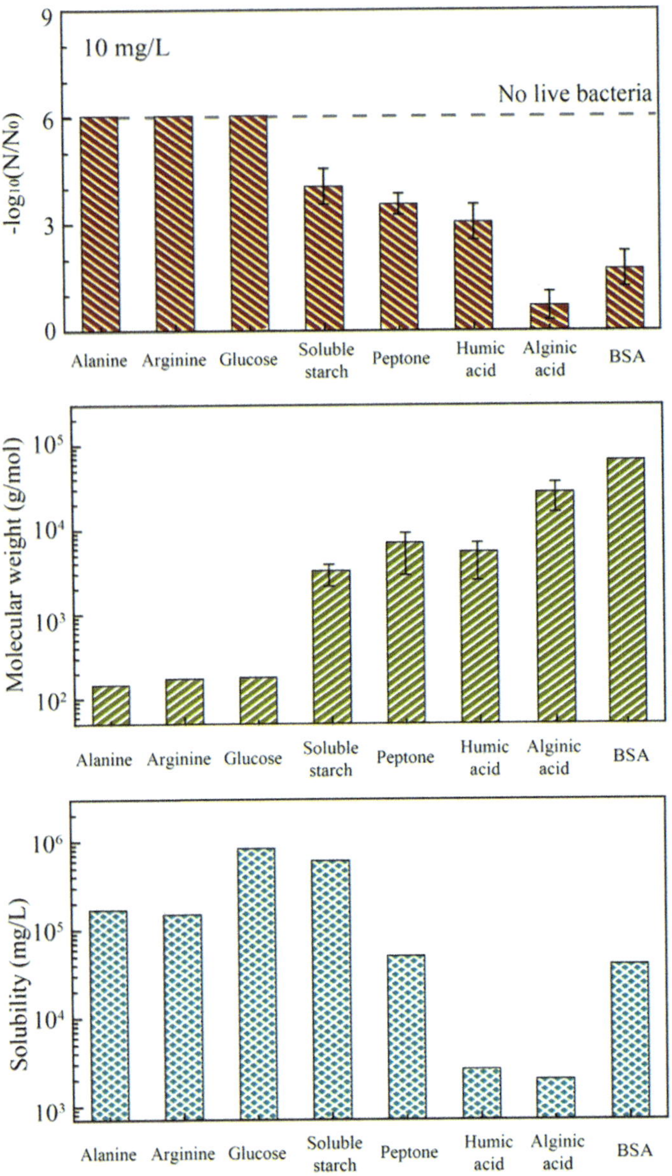

Fig. 7.4 Impact of organic matter properties on DC-powered nanowire-assisted electroporation

significantly impacted only when the molecular weight is >10^4 g/mol. In the case of similar molecular weights, the lower the solubility of the organics, the more significant the effect on the efficiency. The reason for the effect of organic matter may be that large molecules of organic matter in water commonly have certain charges on the surface due to the surface functional groups. Under the influence of positive and negative electrodes, large molecules of organic matter will adhere to the electrode, thus affecting the bacteria to enter the strong electric field area enhanced by nanowires. In addition, for substances with similar molecular weights, the lower the solubility, the easier it is to adhere and deposit on the electrode surface, aggravating the impact on the disinfection efficiency. Notably, alginate and BSA, which have a significant effect on nanowire-assisted electroporation disinfection, are commonly present at low (<1 mg/L) concentrations in the actual reclaimed water. Therefore, DC-powered nanowire-assisted electroporation disinfection can still achieve efficient treatment of the actual reclaimed water[146].

7.3.4 *Effect of Organic Matter on the Disinfection Efficiency of AC-Powered Nanowire-assisted Electroporation*

The study further analyzed the effect of organic matters on the disinfection efficiency when driven by AC power supply. As shown in Fig. 7.5, compared to the significant effect of organic matters with high concentrations on the disinfection efficiency when driven by DC power supply, the effect of disinfection powered by AC was not significant. At an HRT of 10 s, AC-powered nanowire-assisted electroporation achieved complete disinfection (log removal rate > 6.0) in various organic solutions, except for high concentrations of alginate, which had a slight effect on the disinfection efficiency (log removal rate 4.7 log). The reason for the low effect of organic matter on the efficiency of AC-powered electroporation is that in the high-frequency AC-powered mode, the electrode is less attractive to the macromolecular organic matter. These macromolecular organic matters are less likely to deposit on the electrode surface and therefore do not prevent bacteria from entering the strong electric field area near the electrode surface. In summary, the study showed that AC-powered electroporation disinfection was more suitable for the disinfection of reclaimed water with complex water quality.

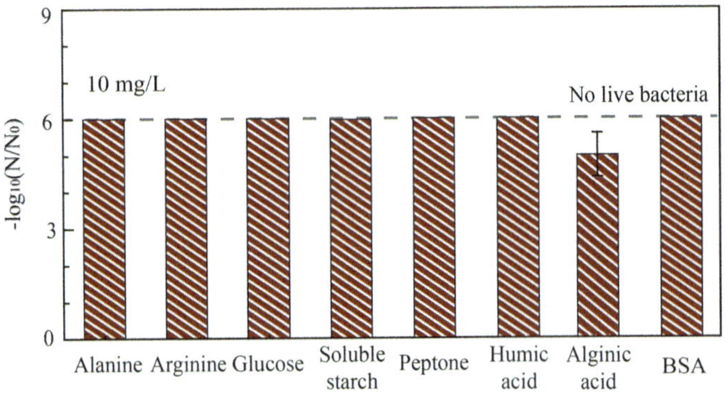

Fig. 7.5 Effect of organic matter on the disinfection efficiency of AC-powered nanowire-assisted electroporation

7.4 Disinfection Performance of Nanowire-assisted Electroporation on Reclaimed Water

7.4.1 Disinfection Performance of Nanowire-assisted Electroporation for Secondary Effluent from Wastewater Reclamation Treatment Plants

To study and analyze the effect of nanowire-assisted electroporation disinfection on actual reclaimed water treatment. Firstly, different cultured pure bacteria were diluted in the secondary effluent after membrane flirtation to determine the disinfection performance. As shown in Fig. 7.6, nanowire-assisted electroporation had a similar disinfection performance on bacteria that diluted in the secondary effluent of four municipal wastewater treatment plants (XJH, BXH, YZ, and QH). The tested disinfection technology achieved complete disinfection (log removal > 6.0) of Gram-negative bacteria (*E. coli* and *E. faecalis*) at 1 V, 5 s operation, while achieved complete disinfection (log removal > 6.0) of Gram-positive bacteria (*B. subtilis* and *E. hormaechei*) at 1 V, 6 s operation. The reason why Gram-positive bacteria were more tolerant to nanowire-assisted electroporation disinfection was due to their thicker cell wall structures. Overall, nanowire-assisted electroporation disinfection achieved complete disinfection for many model bacteria that diluted in the secondary effluent, and this technology proved to have a strong potential for practical application.

Considering the different concentrations of bacteria existing in the actual water bodies, the treatment load of the technology during the disinfection process varies. The study further analyzed the efficiency of nanowire-assisted electroporation disinfection for different concentrations of bacteria in the actual reclaimed water. *E. coli* suspensions were diluted into the secondary effluent of the wastewater treatment plant (XJH) to make final concentrations of *E. coli* as 10^9, 10^7, 10^5, and 10^3 CFU/mL, respectively. As shown in Fig. 7.7, the nanowire-assisted electroporation disinfection

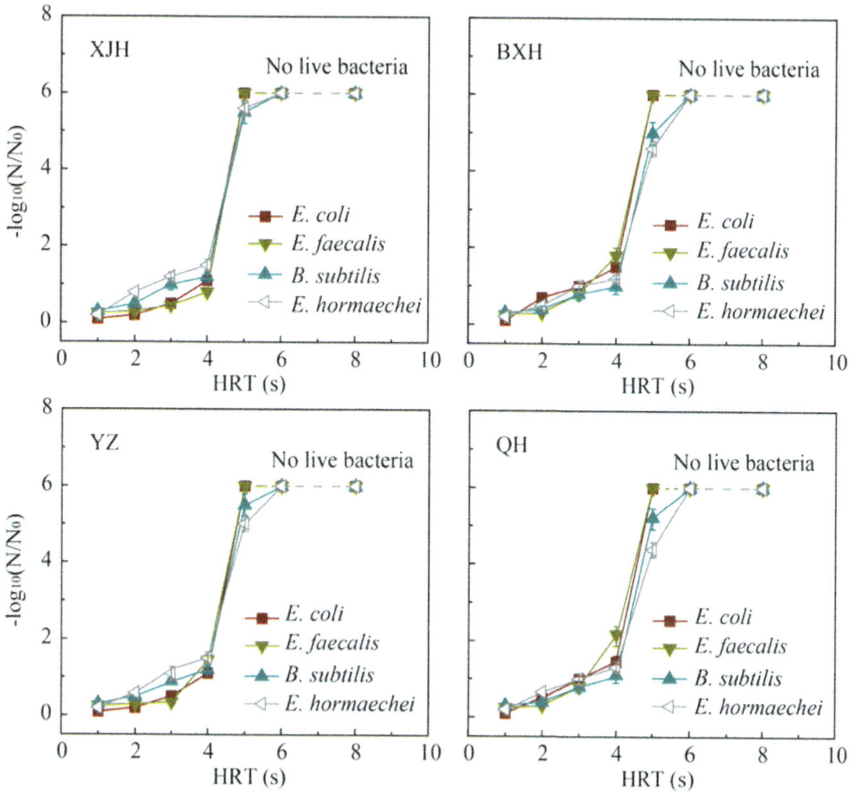

Fig. 7.6 Disinfection efficiency of nanowire-assisted electroporation for different bacteria in secondary effluent from wastewater reclamation treatment plants

achieved complete disinfection of the secondary effluent with different concentrations of *E. coli* when applied with 1 V and 5 s operation. This feature also confirmed the feasibility of nanowire-assisted electroporation disinfection to treat the actual reclaimed water with different biological loads.

The disinfection efficiency for indigenous bacteria in actual wastewater was analyzed using the secondary effluent from four municipal wastewater plants. As shown in Fig. 7.8, the nanowire-assisted electroporation disinfection was highly effective in bacterial inactivation in the actual reclaimed water. At an external voltage of 1 V and an HRT of 6 s, complete disinfection was achieved except for the November 201,610 sample from the HJX WWTP. And for the November 2016 10 sample in the HJX wastewater treatment plant, the nanowire-assisted electroporation disinfection achieved complete disinfection at an HRT of 8 s. This result further proved that nanowire-assisted electroporation disinfection showed great potential to treat the actual reclaimed water.

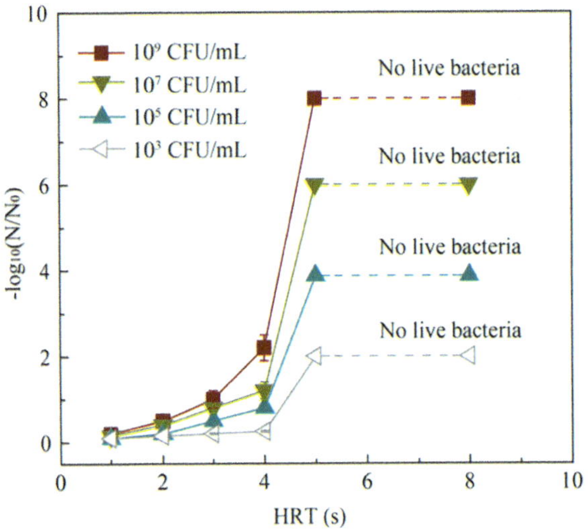

Fig. 7.7 Disinfection efficiency of nanowire-assisted electroporation for different concentrations of *E. coli* in secondary effluent from wastewater reclamation treatment plants

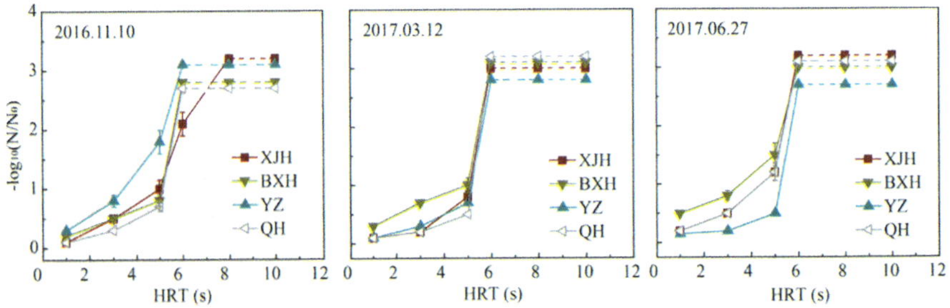

Fig. 7.8 Disinfection efficiency of nanowire-assisted electroporation for indigenous bacteria in secondary effluent from wastewater reclamation treatment plants

7.4.2 Disinfection Performance of Nanowire-assisted Electroporation for the Receiving Water Bodies of Reclaimed Water

Commonly, after treatment, the reclaimed water will be stored in the natural receiving water undergo an eco self-cleaning process to ensure a high water quality. The study further analyzed the effectiveness of nanowire-assisted electroporation disinfection technology on the receiving water bodies of reclaimed water. The Olympic Park water system and Gaobeidian Lake were selected as the typical natural receiving water bodies. Different model bacteria were diluted in the water. As shown in Fig. 7.9,

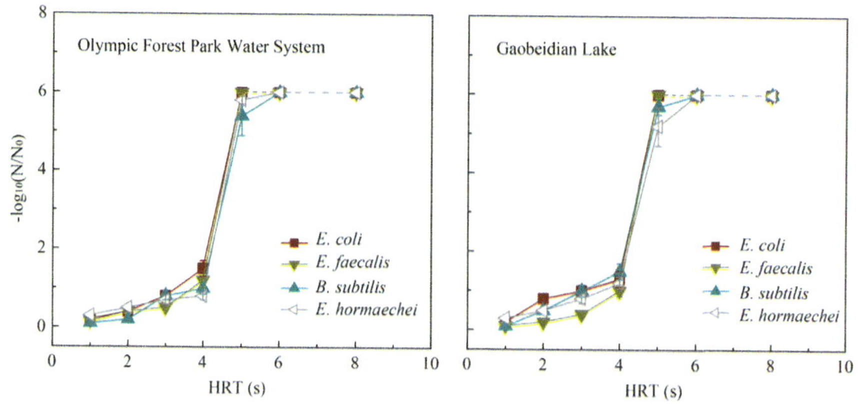

Fig. 7.9 Disinfection efficiency of nanowire-assisted electroporation for model bacteria in receiving water bodies of reclaimed water

nanowire-assisted electroporation disinfection achieved effective bacterial disinfection in the receiving water bodies. For the Olympic Park water system and Gaobeidian lakes, nanowire-assisted electroporation disinfection achieved complete disinfection of Gram-negative bacteria (*E. coli* and *E. faecalis*) at 1 V, 5 s operation (log removal rate > 6.0). For the Gram-positive bacteria (*B. subtilis* and *E. hormaechei*), in the receiving water bodies, nanowire-assisted electroporation disinfection achieved complete disinfection (log removal rate > 6.0) at 1 V, 6 s operation.

The study further analyzed the efficiency of nanowire-assisted electroporation disinfection for the inactivation of bacteria with different concentrations in the receiving water bodies. As shown in Fig. 7.10, nanowire-assisted electroporation disinfection achieved complete disinfection of different concentrations of *E. coli* that diluted in the receiving waters. Log removal of 8.0, 6.0, 4.0, and 2.0 was achieved for receiving waters (Olympic Park water system and Gaobeidian Lake) with bacterial concentrations of 10^9, 10^7, 10^5, and 10^3 CFU/mL, respectively. The nanowire-assisted electroporation disinfection was demonstrated to achieve complete disinfection over a wide range of microbial concentrations. This feature also confirmed the applicability of nanowire-assisted electroporation disinfection for the actual reclaimed water disinfection.

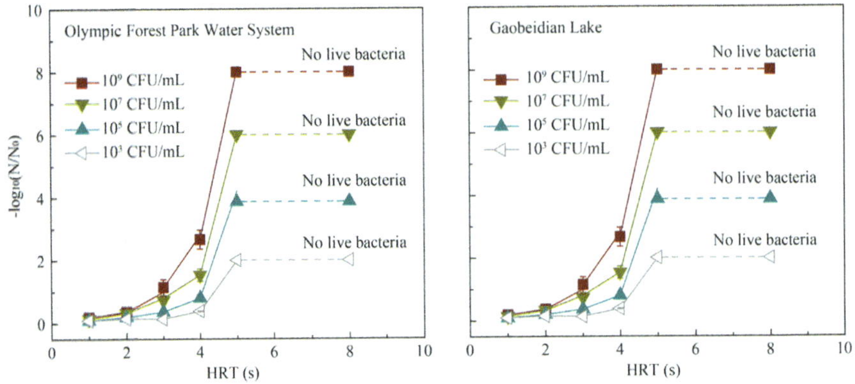

Fig. 7.10 Disinfection efficiency of nanowire-assisted electroporation for different concentrations of *E. coli* in receiving water bodies of reclaimed water

7.5 Summary

(1) The nanowire-assisted electroporation disinfection is able to achieve efficient disinfection of microorganisms in secondary effluent from municipal wastewater treatment plants. The disinfection performance of different model bacteria diluted in secondary effluent from wastewater treatment plants was systematically studied. Nanowire-assisted electroporation disinfection achieved complete disinfection in the secondary effluent of municipal wastewater treatment plants.

(2) Highly efficient disinfection of microorganisms in typical reclaimed water receiving waters can be achieved using nanowire-assisted electroporation disinfection. The disinfection performance of different model bacteria diluted in receiving water bodies of the reclaimed water were systematically studied. Nanowire-assisted electroporation achieved efficient inactivation of microorganisms in reclaimed water receiving waters.

Chapter 8
Conclusions and Perspectives

8.1 Conclusions

In this study, a nanowire-assisted electroporation disinfection system based on the internal filtration treatment mode using porous electrodes was constructed. The nanowire morphology control method during the electrode preparation process was analyzed. The inactivation efficiency of nanowire-assisted electroporation disinfection on the model microbes in reclaimed water and reveal the mechanism of nanowire-assisted electroporation disinfection was investigated. This study further clarified the electroporation reactor optimization method, developed the electrode surface strengthening strategy, and optimized the reactor power supply to improve the electrode lifetime. Finally, this study investigated the effect of nanowire-assisted electroporation disinfection on actual reclaimed water. The following main results were obtained:

(1) The method for the preparation of the CuO nanowire-modified copper foam electrode was established. CuO nanowires with a length of 15 μm and a diameter of 30 nm were grown on the inner surface of the copper foam electrode by using the thermal oxidation method by heating in the air atmosphere at 400 °C. The synthesized nanowires had good morphology and grew uniformly along the direction perpendicular to the inner surface of the electrode.
(2) The nanowire-assisted electroporation disinfection device based on the internal filtration treatment mode using porous electrodes was constructed, and the disinfection efficiency was investigated. Complete disinfection of the model bacteria and bacteria diluted in the secondary effluent from the wastewater treatment plant was achieved with 1 V, 10 s operation. The absence of chlorine generation in the disinfected effluent ensured a very low potential for chlorine-containing disinfection by-products during the treatment process, while the extremely low current in the circuit ensured a high energy utilization efficiency for electroporation disinfection.

Z. Huo, *Nanowire-assisted Flow-through Electrode Enabling Electroporation Disinfection of Reclaimed Water*, Springer Theses,
https://doi.org/10.1007/978-981-99-4502-3_8

(3) To clarify the change of live bacteria concentration during the storage process after disinfection, inactivated bacteria after nanowire-assisted electroporation treatment showed no reactivation or regrowth. For the bacteria undergoing insufficient treatment that did not inactivate immediately after operation, they were inactivated gradually during the subsequent storage process. The mechanism of inactivation was confirmed that the electroporation process damaged the cell membrane structure, and the cell membrane damage continued to intensify during the storage process, which eventually led to bacterial inactivation. It was demonstrated that electroporation disinfection not only ensured that the bacteria showed no reactivation or regrowth after inactivation but also promoted the gradual inactivation of the treated live bacteria during storage.

(4) Clarify the nanowire morphology optimization strategy for the nanowire-modified electrodes. The nanowire aspect ratio (length: diameter) was the main factor affecting the disinfection effect. The nanowire-modified electrode achieved complete disinfection when the diameter of the nanowire was on the nanometer scale and the aspect ratio was higher than 40. In addition, the electrode pore size was another main electrode design factor affecting electroporation disinfection, and the positive electrode had a higher contribution to disinfection during actual operation.

(5) The method for fabricating the Cu_3P nanowire-modified copper foam electrode was established. Cu_3P nanowires with a length of 20 μm and a diameter of 400 nm were able to fabricate uniformly on the surface of copper foam using an electrochemical method and phosphorylation method. The Cu_3P nanowires showed high physical strength and chemical resistance which effectively resisted the scouring of water in the disinfection process.

(6) The use of Cu_3P nanowire-modified copper foam electrodes achieved efficient inactivation of the model microbes in DI water and the actual secondary effluent. The disinfection mechanism was confirmed to be electroporation. The lifetime of the electrode was significantly increased due to the strong physical stability and chemical resistance of the Cu_3P nanowires. The study further confirmed the electrochemical reaction at the solution-electrode interface as the main mechanism of electrode loss during long-term operation.

(7) Applying thin films to uniformly cover the electrodes effectively improved the electrode lifetime. By using PDA as a protective film and wrapping the PDA film uniformly around the nanowire electrode surface, the lifetime of the electrode was significantly increased while ensuring efficient electroporation disinfection efficiency. In addition, electrode loss was significantly reduced during long-term operation after covered with PDA film, and electrochemical reactions were still the main cause of electrode loss.

(8) Long-term electroporation disinfection was confirmed feasible driven by high-frequency AC power supply. The preferred AC frequency was 10^6 Hz, which prevented electrochemical reactions while ensured efficient nanowire-assisted electroporation disinfection. The effect of electrochemical reactions on electrode loss was fundamentally eliminated. When powered by high-frequency AC, the PDA-coated Cu_3P nanowire-modified copper foam electrode achieved

complete disinfection for 15 consecutive days with tract electrode loss (total copper concentration releasing from the electrode < 4 μg/L in the effluent).

(9) Nanowire-assisted electroporation disinfection achieved efficient disinfection of actual reclaimed water. Both inorganic factors (pH, ionic strength, and ionic type) and organic matter showed trace effects on electroporation disinfection. Nanowire-assisted electroporation disinfection also enabled complete inactivation of bacteria in secondary effluent from municipal wastewater treatment plants. In addition, nanowire-assisted electroporation disinfection was effective for different biological loads in the actual wastewater reclamation process.

8.2 Perspectives

Based on the findings of the existing studies, it is recommended that the following further studies be carried out.

(1) Research on nanowire-assisted electroporation disinfection using carbon, silicon, and titanium materials as substrates to guarantee electrode stability and reduce toxicity of lost elements is necessary. Further research on electrode surface strengthening technology can use atomic layer deposition technology and magnetron sputtering technology to uniformly coat an inert protective film (e.g., Al_2O_3) on the surface of nanowire electrodes to enhance their lifetime.

(2) Research on nanowire-assisted electroporation disinfection for more indicator model microorganisms (e.g., viruses, giardia, and cryptosporidium) and high-risk microorganisms (e.g., antibiotic-resistant bacteria) in reclaimed water is necessary. Further research also needs to expand the application scope of nanowire-assisted electroporation disinfection technology to industrial and medical wastewater treatment.

(3) Research on reactor scale-up and establish pilot or actual scale disinfection treatment is needed. Further research also needs to evaluate the economic benefits of nanowire-assisted electroporation disinfection technology based on large-scale practical applications and compare it with existing reclaimed water disinfection technologies. Applying novel power sources (e.g., alkaline battery, solar cell, hand generator, and nanogenerator) to make nanowire-assisted electroporation disinfection applicable to actual treatment applications for different scenarios.

Curriculum Vitae

Dr. Zheng-Yang Huo is an assistant professor at the School of Environment, Renmin University of China, since 2021. He received his Ph.D. from the School of Environment, Tsinghua in 2019 under the supervision of Prof. Hong-Ying Hu, and worked as a research professor with Prof. Kim Sang-Woo at Sungkyunkwan University, funded by the Korea Research Fellowship. Dr Huo focused on novel materials for water disinfection and biological pollution control in the environment. He has completed six projects in the past five years, including the National Key R&D Program Youth Project (Task Leader), the National Natural Science Foundation Youth Project, and the European Union Marie Curie Scholars Research Fund. Dr Huo has published more than 30 papers in journals such as Nature Communication, Advanced Energy Materials, Environmental Science & Technology, Nano Letter, Water Research, etc., with more than 1200 citations. He is currently a young editorial board member of several journals such as Frontiers of Environmental Science & Engineering. Dr Huo was selected for the Beijing Association of Science and Technology Young Talent Support Project (2023).

Z. Huo, *Nanowire-assisted Flow-through Electrode Enabling Electroporation Disinfection of Reclaimed Water*, Springer Theses,
https://doi.org/10.1007/978-981-99-4502-3

References

1. The United Nations Children's Fund and World Health Organization (2019) 1 in 3 people globally do not have access to safe drinking water [J]
2. FY Ramírez-Castillo A Loera-Muro M Jacques 2015 Waterborne pathogens: detection methods and challenges[J] Pathogens 4 2 307 334
3. PK Pandey PH Kass ML Soupir 2014 Contamination of water resources by pathogenic bacteria[J] AMB Express 4 1 1 16
4. A Malik A Yasar AB Tabinda 2012 Water-borne diseases, cost of illness and willingness to pay for diseases interventions in rural communities of developing countries[J] Iran J Public Health 41 6 39
5. SA Bartels MJ VanRooyen 2012 Medical complications associated with earthquakes[J] The Lancet 379 9817 748 757
6. C He CP Harden Y Liu 2020 Comparison of water resources management between China and the United States[J] Geog Sustain 1 2 98 108
7. Qu J, Zhao J, Ren N et al (2017) Key basic scientific issues in Urban wastewater regeneration and recycling[J]. China Basic Sci 19(1):6–12. 曲久辉, 赵进才, 任南琪, 等. 城市污水再生与循环利用的关键基础科学问题[J]. 中国基础科学, 2017, 19(1): 6–12
8. Ministry of Water Resources of the People's Republic of China (2016) China Water Resources Bulletin[EB], (4). 中华人民共和国水利部. 中国水资源公报[EB]. 2016, (4)
9. Hu H, Shi L, Xu C, et al (2015) Regional water resource recycling model: concept, structure, and characteristics[J]. Environ Sci Res 28(006):839–847. 胡洪营, 石磊, 许春华, 等. 区域水资源介循环利用模式: 概念· 结构· 特征[J]. 环境科学研究, 2015, 28(006): 839–847
10. H Lee TP Tan 2016 Singapore's experience with reclaimed water: NEWater[J] Int J Water Resour Dev 32 4 611 621
11. AS Stillwell CW King ME Webber 2011 The energy-water nexus in Texas[J] Ecol Soc 16 1 2
12. RIL Eggen J Hollender A Joss 2014 Reducing the discharge of micropollutants in the aquatic environment: the benefits of upgrading wastewater treatment plants[J] Environ Sci Technol 48 14 7683 7689
13. L Yi W Jiao X Chen 2011 An overview of reclaimed water reuse in China[J] J Environ Sci 23 10 1585 1593
14. S Lyu W Chen W Zhang 2016 Wastewater reclamation and reuse in China: opportunities and challenges[J] J Environ Sci 39 86 96
15. J Qu H Wang K Wang 2019 Municipal wastewater treatment in China: development history and future perspectives[J] Front Environ Sci Eng 13 6 1 7
16. D Courault I Albert S Perelle 2017 Assessment and risk modeling of airborne enteric viruses emitted from wastewater reused for irrigation[J] Sci Total Environ 592 512 526

Z. Huo, *Nanowire-assisted Flow-through Electrode Enabling Electroporation Disinfection of Reclaimed Water*, Springer Theses,
https://doi.org/10.1007/978-981-99-4502-3

17. L Cai T Zhang 2013 Detecting human bacterial pathogens in wastewater treatment plants by a high-throughput shotgun sequencing technique[J] Environ Sci Technol 47 10 5433 5441
18. L Yi W Jiao X Chen 2011 An overview of reclaimed water reuse in China[J] J Environ Sci 23 10 1585 1593
19. JJ Huang HY Hu SQ Lu 2012 Monitoring and evaluation of antibiotic-resistant bacteria at a municipal wastewater treatment plant in China[J] Environ Int 42 31 36
20. AH Havelaar TJ Nieuwstad 1985 Bacteriophages and fecal bacteria as indicators of chlorination efficiency of biologically treated wastewater[J] Water Poll Control Feder 57 11 1084 1088
21. J Pépin N Saheb MA Coulombe 2005 Emergence of fluoroquinolones as the predominant risk factor for Clostridium difficile–associated diarrhea: a cohort study during an epidemic in Quebec[J] Clin Infect Dis 41 9 1254 1260
22. E Espigares A Bueno M Espigares 2006 Isolation of Salmonella serotypes in wastewater and effluent: effect of treatment and potential risk[J] Int J Hyg Environ Health 209 1 103 107
23. JG Songer 1996 Clostridial enteric diseases of domestic animals[J] Clin Microbiol Rev 9 2 216
24. ME Nyenje CE Odjadjare NF Tanih 2012 Foodborne pathogens recovered from ready-to-eat foods from roadside cafeterias and retail outlets in Alice, Eastern Cape Province, South Africa: public health implications[J] Int J Environ Res Public Health 9 8 2608 2619
25. I Tabain S Ljubin-Sternak J Cepin-Bogovic 2012 Adenovirus respiratory infections in hospitalized children: clinical findings in relation to species and serotypes[J] Pediatr Infect Dis J 31 7 680 684
26. DP Zheng T Ando RL Fankhauser 2006 Norovirus classification and proposed strain nomenclature[J] Virology 346 2 312 323
27. Widdowson MA, Steele D, Vojdani J, et al (2009) Global rotavirus surveillance: determining the need and measuring the impact of rotavirus vaccines[J], 200(Supplement_1):S1–S8
28. T Solomon P Lewthwaite D Perera 2010 Virology, epidemiology, pathogenesis, and control of enterovirus 71[J] Lancet Infect Dis 10 11 778 790
29. RD Adam 2001 Biology of Giardia lamblia[J] Clin Microbiol Rev 14 3 447 475
30. JB Rose LJ Dickson SR Farrah 1996 Removal of pathogenic and indicator microorganisms by a full-scale water reclamation facility[J] Water Res 30 11 2785 2797
31. H Wang C Hu X Hu 2012 Effects of disinfectant and biofilm on the corrosion of cast iron pipes in a reclaimed water distribution system[J] Water Res 46 4 1070 1078
32. H Huang QY Wu HY Hu 2012 Dichloroacetonitrile and dichloroacetamide can form independently during chlorination and chloramination of drinking waters, model organic matters, and wastewater effluents[J] Environ Sci Technol 46 19 10624 10631
33. X Tang QY Wu Y Du 2014 Anti-estrogenic activity formation potential assessment and precursor analysis in reclaimed water during chlorination[J] Water Res 48 1 490 497
34. DL Sedlak U Gunten von 2011 The chlorine dilemma[J] Science 331 6013 42 43
35. M Taran-Benshoshan N Ofer VO Dalit 2015 Cryptosporidium and Giardia removal by secondary and tertiary wastewater treatment[J] J Environ Sci Health, Part A 50 12 1265 1273
36. M Guo J Huang H Hu 2012 UV inactivation and characteristics after photoreactivation of Escherichia coli with plasmid: health safety concern about UV disinfection[J] Water Res 46 13 4031 4036
37. MT Guo QB Yuan J Yang 2013 Microbial selectivity of UV treatment on antibiotic-resistant heterotrophic bacteria in secondary effluents of a municipal wastewater treatment plant[J] Water Res 47 16 6388 6394
38. K Oguma H Katayama H Mitani 2001 Determination of pyrimidine dimers in Escherichia coli and Cryptosporidium parvum during UV light inactivation, photoreactivation, and dark repair[J] Appl Environ Microbiol 67 10 4630 4637
39. DS Francy EA Stelzer RN Bushon 2012 Comparative effectiveness of membrane bioreactors, conventional secondary treatment, and chlorine and UV disinfection to remove microorganisms from municipal wastewaters[J] Water Res 46 13 4164 4178

40. FJ Simmons I Xagoraraki 2011 Release of infectious human enteric viruses by full-scale wastewater utilities[J] Water Res 45 12 3590 3598
41. CH Jeong C Postigo SD Richardson 2015 Occurrence and comparative toxicity of haloacetaldehyde disinfection byproducts in drinking water[J] Environ Sci Technol 49 23 13749 13759
42. C Li D Wang X Xu 2017 Formation of known and unknown disinfection by-products from natural organic matter fractions during chlorination, chloramination, and ozonation[J] Sci Total Environ 587 177 184
43. X Zhao HY Hu T Yu 2014 Effect of different molecular weight organic components on the increase of microbial growth potential of secondary effluent by ozonation[J] J Environ Sci 26 11 2190 2197
44. V Lazarova P Savoye ML Janex 1999 Advanced wastewater disinfection technologies: state of the art and perspectives[J] Water Sci Technol 40 4–5 203 213
45. R Wang S Guan A Sato 2013 Nanofibrous microfiltration membranes capable of removing bacteria, viruses and heavy metal ions[J] J Membr Sci 446 376 382
46. H Ma BS Hsiao B Chu 2014 Functionalized electrospun nanofibrous microfiltration membranes for removal of bacteria and viruses[J] J Membr Sci 452 446 452
47. JC Weaver YA Chizmadzhev 1996 Theory of electroporation: a review[J] Bioelectrochem Bioenerg 41 2 135 160
48. H Aihara J Miyazaki 1998 Gene transfer into muscle by electroporation in vivo[J] Nat Biotechnol 16 9 867
49. Dower WJ, Miller JF, Ragsdale CW (1988) High efficiency transformation of *E. coli* by high voltage electroporation[J]. Nucleic Acids Res 16(13):6127–6145
50. ML Yarmush A Golberg G Serša 2014 Electroporation-based technologies for medicine: principles, applications, and challenges[J] Annu Rev Biomed Eng 16 295 320
51. NY Sardesai DB Weiner 2011 Electroporation delivery of DNA vaccines: prospects for success[J] Curr Opin Immunol 23 3 421 429
52. L Wang L Chierico D Little 2012 Encapsulation of biomacromolecules within polymersomes by electroporation[J] Angew Chem Int Ed 51 44 11122 11125
53. C Zhang H Hu 2014 High-efficiency nuclear transformation of the diatom Phaeodactylum tricornutum by electroporation[J] Mar Genomics 16 63 66
54. JH Wang W Hung SH Tsai 2011 High efficiency transformation by electroporation of Yarrowia lipolytica[J] J Microbiol 49 3 469 472
55. TP Kingham AM Karkar MI D'Angelica 2012 Ablation of perivascular hepatic malignant tumors with irreversible electroporation[J] J Am Coll Surg 215 3 379 387
56. KR Thomson W Cheung SJ Ellis 2011 Investigation of the safety of irreversible electroporation in humans[J] J Vasc Interv Radiol 22 5 611 621
57. C Jiang RV Davalos JC Bischof 2015 A review of basic to clinical studies of irreversible electroporation therapy[J] IEEE Trans Biomed Eng 62 1 4 20
58. RCG Martin II K McFarland S Ellis 2012 Irreversible electroporation therapy in the management of locally advanced pancreatic adenocarcinoma[J] J Am Coll Surg 215 3 361 369
59. C Liu X Xie W Zhao 2013 Conducting nanosponge electroporation for affordable and high-efficiency disinfection of bacteria and viruses in water[J] Nano Lett 13 9 4288 4293
60. T Kotnik W Frey M Sack 2015 Electroporation-based applications in biotechnology[J] Trends Biotechnol 33 8 480 488
61. NJ Rowan SJ Macgregor JG Anderson 2000 Pulsed electric field inactivation of diarrhoeagenic Bacillus cereus through irreversible electroporation[J] Lett Appl Microbiol 31 2 110 114
62. CA Gusbeth W Frey T Schwartz 2009 Critical comparison between the pulsed electric field and thermal decontamination methods of hospital wastewater[J] Acta Phys Polonica-Series A Gen Phy 115 6 1092
63. A Rieder T Schwartz K Schön-Hölz 2008 Molecular monitoring of inactivation efficiencies of bacteria during pulsed electric field treatment of clinical wastewater[J] J Appl Microbiol 105 6 2035 2045

64. G Saulis 2010 Electroporation of cell membranes: the fundamental effects of pulsed electric fields in food processing[J] Food Eng Rev 2 2 52 73
65. C Gusbeth W Frey H Volkmann 2009 Pulsed electric field treatment for bacteria reduction and its impact on hospital wastewater[J] Chemosphere 75 2 228 233
66. CN Haas D Aturaliye 1999 Semi-quantitative characterization of electroporation-assisted disinfection processes for inactivation of Giardia and Cryptosporidium[J] J Appl Microbiol 86 6 899 905
67. CN Haas DN Aturaliye 1999 Kinetics of electroporation-assisted chlorination of Giardia muris[J] Water Res 33 8 1761 1766
68. BC Hodges EL Cates JH Kim 2018 Challenges and prospects of advanced oxidation water treatment processes using catalytic nanomaterials[J] Nat Nanotechnol 13 8 642
69. JLC Santos V Geraldes S Velizarov 2010 Characterization of fluid dynamics and mass-transfer in an electrochemical oxidation cell by experimental and CFD studies[J] Chem Eng J 157 2–3 379 392
70. M Law LE Greene JC Johnson 2005 Nanowire dye-sensitized solar cells[J] Nat Mater 4 6 455
71. X Duan Y Huang R Agarwal 2003 Single-nanowire electrically driven lasers[J] Nature 421 6920 241
72. Y Cui Q Wei H Park 2001 Nanowire nanosensors for highly sensitive and selective detection of biological and chemical species[J] Science 293 5533 1289 1292
73. S Xu Y Qin C Xu 2010 Self-powered nanowire devices[J] Nat Nanotechnol 5 5 366
74. ZL Wang J Song 2006 Piezoelectric nanogenerators based on zinc oxide nanowire arrays[J] Science 312 5771 242 246
75. F Patolsky G Zheng CM Lieber 2006 Nanowire-based biosensors[J] Anal Chem 78 13 4260 4269
76. L Billot ML Chapelle de La AS Grimault 2006 Surface enhanced Raman scattering on gold nanowire arrays: evidence of strong multipolar surface plasmon resonance enhancement[J] Chem Phys Lett 422 4–6 303 307
77. C Liu X Xie W Zhao 2014 Static electricity powered copper oxide nanowire microbicidal electroporation for water disinfection[J] Nano Lett 14 10 5603 5608
78. Huo ZY, Kim YJ, Suh IY et al. (2021) Triboelectrification induced self-powered microbial disinfection using nanowire-enhanced localized electric field[J]. Nat Commun
79. DT Schoen AP Schoen L Hu 2010 High speed water sterilization using one-dimensional nanostructures[J] Nano Lett 10 9 3628 3632
80. G Wang H Feng L Hu 2018 An antibacterial platform based on capacitive carbon-doped TiO_2 nanotubes after direct or alternating current charging[J] Nat Commun 9 1 2055
81. C Liu X Xie W Zhao 2014 Static electricity powered copper oxide nanowire microbicidal electroporation for water disinfection[J] Nano Lett 14 10 5603 5608
82. J Tian H Feng L Yan 2017 A self-powered sterilization system with both instant and sustainable anti-bacterial ability[J] Nano Energy 36 241 249
83. J Radjenovic DL Sedlak 2015 Challenges and opportunities for electrochemical processes as next-generation technologies for the treatment of contaminated water[J] Environ Sci Technol 49 19 11292 11302
84. ZY Huo Y Du Z Chen 2020 Evaluation and prospects of nanomaterial-enabled innovative processes and devices for water disinfection: a state-of-the-art review[J] Water Res 173 115581
85. L Ma M Zhou G Ren 2016 A highly energy-efficient flow-through electro-Fenton process for organic pollutants degradation[J] Electrochim Acta 200 222 230
86. RM Darling KG Gallagher JA Kowalski 2014 Pathways to low-cost electrochemical energy storage: a comparison of aqueous and nonaqueous flow batteries[J] Energy Environ Sci 7 11 3459 3477
87. Z Gu X An H Lan 2019 Microfluidic-enhanced 3-D photoanodes with free interfacial energy barrier for photoelectrochemical applications[J] Appl Catal B 244 740 747
88. H Liu XY Ni ZY Huo 2019 Carbon fiber-based flow-through electrode system (FES) for water disinfection via direct oxidation mechanism with a sequential reduction–oxidation process[J] Environ Sci Technol 53 6 3238 3249

89. L Guo K Ding K Rockne 2016 Bacteria inactivation at a sub-stoichiometric titanium dioxide reactive electrochemical membrane[J] J Hazard Mater 319 137 146
90. Á Anglada AM Urtiaga I Ortiz 2010 Laboratory and pilot plant scale study on the electrochemical oxidation of landfill leachate[J] J Hazard Mater 181 1–3 729 735
91. ZY Huo X Xie T Yu 2016 Nanowire-modified three-dimensional electrode enabling low-voltage electroporation for water disinfection[J] Environ Sci Technol 50 14 7641 7649
92. H Wang ZJ Ren 2013 A comprehensive review of microbial electrochemical systems as a platform technology[J] Biotechnol Adv 31 8 1796 1807
93. H Pan X Yin J Xue 2016 In-situ synthesis of hierarchically porous and polycrystalline carbon nanowires with excellent microwave absorption performance[J] Carbon 107 36 45
94. B Weng W Wei H Wu 2016 Bifunctional CoP and CoN porous nanocatalysts derived from ZIF-67 in situ grown on nanowire photoelectrodes for efficient photoelectrochemical water splitting and CO2 reduction[J] J Mater Chem A 4 40 15353 15360
95. SM Jung DJ Preston HY Jung 2016 Porous Cu nanowire aerosponges from one-step assembly and their applications in heat dissipation[J] Adv Mater 28 7 1413 1419
96. XL Zhou YK Zeng XB Zhu 2016 A high-performance dual-scale porous electrode for vanadium redox flow batteries[J] J Power Sou 325 329 336
97. XY Yu Y Feng B Guan 2016 Carbon coated porous nickel phosphides nanoplates for highly efficient oxygen evolution reaction[J] Energy Environ Sci 9 4 1246 1250
98. S Kitagawa R Kitaura S Noro 2004 Functional porous coordination polymers[J] Angew Chem Int Ed 43 18 2334 2375
99. ME Davis 2002 Ordered porous materials for emerging applications[J] Nature 417 6891 813
100. L Sun MG Campbell M Dincă 2016 Electrically conductive porous metal–organic frameworks[J] Angew Chem Int Ed 55 11 3566 3579
101. C Wang T Lin N Li 2016 Heat transfer enhancement of phase change composite material: copper foam/paraffin[J] Renew Energy 96 960 965
102. X Gao J Zhou R Du 2016 Robust superhydrophobic foam: a graphdiyne-based hierarchical architecture for oil/water separation[J] Adv Mater 28 1 168 173
103. X Jiang T Herricks Y Xia 2002 CuO nanowires can be synthesized by heating copper substrates in air [J] Nano Lett 2 12 1333 1338
104. L Liao Z Zhang B Yan 2009 Multifunctional CuO nanowire devices: p-type field effect transistors and CO gas sensors[J] Nanotechnology 20 8 085203
105. YS Kim IS Hwang SJ Kim 2008 CuO nanowire gas sensors for air quality control in automotive cabin[J] Sens Actuators, B Chem 135 1 298 303
106. X Zhao P Wang Z Yan 2015 Room temperature photoluminescence properties of CuO nanowire arrays[J] Opt Mater 42 544 547
107. S Ma Z Peng AH Kitai 2018 A CuO Nanowire-based alternating current oxide powder electroluminescent device with high stability[J] Angew Chem Int Ed 57 35 11267 11272
108. CM Tang YB Wang RH Yao 2016 Enhanced adhesion and field emission of CuO nanowires synthesized by simply modified thermal oxidation technique[J] Nanotechnology 27 39 395605
109. J Tian Q Liu N Cheng 2014 Self-supported Cu3P nanowire arrays as an integrated high-performance three-dimensional cathode for generating hydrogen from water[J] Angew Chem Int Ed 53 36 9577 9581
110. H Hou Y Xie Q Li 2005 Large-scale synthesis of single-crystalline quasi-aligned submicrometer CuO ribbons[J] Cryst Growth Des 5 1 201 205
111. Z Zhang R Dua L Zhang 2013 Carbon-layer-protected cuprous oxide nanowire arrays for efficient water reduction[J] ACS Nano 7 2 1709 1717
112. M Fan Y Chen Y Xie 2016 Half-cell and full-cell applications of highly stable and binder-free sodium ion batteries based on Cu3P nanowire anodes[J] Adv Func Mater 26 28 5019 5027
113. AA Dubale WN Su AG Tamirat 2014 The synergetic effect of graphene on Cu2O nanowire arrays as a highly efficient hydrogen evolution photocathode in water splitting[J] J Mater Chem A 2 43 18383 18397
114. H Du X Zhang Q Tan 2017 A Cu3P-CoP hybrid nanowire array: a superior electrocatalyst for acidic hydrogen evolution reactions[J] Chem Commun 53 88 12012 12015

115. JA Rojas-Chapana MA Correa-Duarte Z Ren 2004 Enhanced introduction of gold nanoparticles into vital acidothiobacillus ferrooxidans by carbon nanotube-based microwave electroporation[J] Nano Lett 4 5 985 988
116. M Kaur KP Muthe SK Despande 2006 Growth and branching of CuO nanowires by thermal oxidation of copper[J] J Cryst Growth 289 2 670 675
117. L Yuan Y Wang R Mema 2011 Driving force and growth mechanism for spontaneous oxide nanowire formation during the thermal oxidation of metals[J] Acta Mater 59 6 2491 2500
118. R Mema L Yuan Q Du 2011 Effect of surface stresses on CuO nanowire growth in the thermal oxidation of copper[J] Chem Phys Lett 512 1–3 87 91
119. S Steinhauer E Brunet T Maier 2013 Gas sensing properties of novel CuO nanowire devices[J] Sens Actuators, B Chem 187 50 57
120. S Steinhauer E Brunet T Maier 2013 Suspended CuO nanowires for ppb level H2S sensing in dry and humid atmosphere[J] Sens Actuators, B Chem 186 550 556
121. HT Nguyen F Yao MR Zamfir 2011 Highly Interconnected Si Nanowires for Improved Stability Li-Ion Battery Anode[J] Adv Energy Mater 1 6 1154 1161
122. Y Yang YF Cheng 2011 Electrolytic deposition of Ni-Co-SiC nano-coating for erosion-enhanced corrosion of carbon steel pipes in oilsand slurry[J] Surf Coat Technol 205 10 3198 3204
123. YJ Hwang C Hahn B Liu 2012 Photoelectrochemical properties of TiO2 nanowire arrays: a study of the dependence on length and atomic layer deposition coating[J] ACS Nano 6 6 5060 5069
124. S Sim P Oh S Park 2013 Critical thickness of SiO2 coating layer on Core@ Shell Bulk@ nanowire Si anode materials for Li-Ion batteries[J] Adv Mater 25 32 4498 4503
125. K Evanoff J Benson M Schauer 2012 Ultra strong silicon-coated carbon nanotube nonwoven fabric as a multifunctional lithium-ion battery anode[J] ACS Nano 6 11 9837 9845
126. J Ryu CB Park 2009 Synthesis of diphenylalanine/polyaniline core/shell conducting nanowires by peptide self-assembly[J] Angew Chem 121 26 4914 4917
127. Z Zhang J Zhang B Zhang 2013 Mussel-inspired functionalization of graphene for synthesizing Ag-polydopamine-graphene nanosheets as antibacterial materials[J] Nanoscale 5 1 118 123
128. SH Yoo S Park 2007 Platinum-coated, nanoporous gold nanorod arrays: synthesis and characterization[J] Adv Mater 19 12 1612 1615
129. X Liu J Cao H Li 2013 Mussel-inspired polydopamine: a biocompatible and ultrastable coating for nanoparticles in vivo[J] ACS Nano 7 10 9384 9395
130. J Park TF Brust HJ Lee 2014 Polydopamine-based simple and versatile surface modification of polymeric nano drug carriers[J] ACS Nano 8 4 3347 3356
131. MH Ryou YM Lee JK Park 2011 Mussel-inspired polydopamine-treated polyethylene separators for high-power Li-ion batteries[J] Adv Mater 23 27 3066 3070
132. J Ryu SH Ku H Lee 2010 Mussel-inspired polydopamine coating as a universal route to hydroxyapatite crystallization[J] Adv Func Mater 20 13 2132 2139
133. H Wu J Kong X Yao 2015 Polydopamine-assisted attachment of β-cyclodextrin on porous electrospun fibers for water purification under highly basic condition[J] Chem Eng J 270 101 109
134. KJ Huang DJ Niu X Liu 2011 Direct electrochemistry of catalase at amine-functionalized graphene/gold nanoparticles composite film for hydrogen peroxide sensor[J] Electrochim Acta 56 7 2947 2953
135. J Xu F Shang JHT Luong 2010 Direct electrochemistry of horseradish peroxidase immobilized on a monolayer modified nanowire array electrode[J] Biosens Bioelectron 25 6 1313 1318
136. NM Calvin PC Hanawalt 1988 High-efficiency transformation of bacterial cells by electroporation[J] J Bacteriol 170 6 2796 2801
137. K Kinosita Jr I Ashikawa N Saita 1988 Electroporation of cell membrane visualized under a pulsed-laser fluorescence microscope[J] Biophys J 53 6 1015 1019
138. MS Rahaman CD Vecitis M Elimelech 2012 Electrochemical carbon-nanotube filter performance toward virus removal and inactivation in the presence of natural organic matter[J] Environ Sci Technol 46 3 1556 1564

139. K Wan Z Yu X Li 2015 pH effect on electrochemistry of nitrogen-doped carbon catalyst for oxygen reduction reaction[J] ACS Catal 5 7 4325 4332
140. Gumy D, Morais C, Bowen P, et al (2006) Catalytic activity of commercial of TiO2 powders for the abatement of the bacteria (E. coli) under solar simulated light: influence of the isoelectric point[J]. Appl Catal B: Environ 63(1–2):76–84
141. HHM Rijnaarts W Norde J Lyklema 1995 The isoelectric point of bacteria as an indicator for the presence of cell surface polymers that inhibit adhesion[J] Colloids Surf, B 4 4 191 197
142. FC Tenover 2006 Mechanisms of antimicrobial resistance in bacteria[J] Am J Med 119 6 S3 S10
143. O González A Justo J Bacardit 2013 Characterization and fate of effluent organic matter treated with UV/H2O2 and ozonation[J] Chem Eng J 226 402 408
144. X Shi C Xu H Hu 2016 Characterization of dissolved organic matter in the secondary effluent of pulp and paper mill wastewater before and after coagulation treatment[J] Water Sci Technol 74 6 1346 1353
145. N Her G Amy D McKnight 2003 Characterization of DOM as a function of MW by fluorescence EEM and HPLC-SEC using UVA, DOC, and fluorescence detection[J] Water Res 37 17 4295 4303
146. HY Hu Y Du QY Wu 2016 Differences in dissolved organic matter between reclaimed water source and drinking water source[J] Sci Total Environ 551 133 142
147. RV Davalos LM Mir B Rubinsky 2005 Tissue ablation with irreversible electroporation[J] Ann Biomed Eng 33 2 223
148. Guzman M, Dille J, Godet S (2012) Synthesis and antibacterial activity of silver nanoparticles against gram-positive and gram-negative bacteria[J]. Nanomed Nanotechn Biol Med 8(1):37–45
149. YC Pang JY Xi Y Xu 2016 Shifts of live bacterial community in secondary effluent by chlorine disinfection revealed by Miseq high-throughput sequencing combined with propidium monoazide treatment[J] Appl Microbiol Biotechnol 100 14 6435 6446
150. J Pérez-Juste LM Liz-Marzán S Carnie 2004 Electric-field-directed growth of gold nanorods in aqueous surfactant solutions[J] Adv Func Mater 14 6 571 579
151. M Liu Y Pang B Zhang 2016 Enhanced electrocatalytic CO_2 reduction via field-induced reagent concentration[J] Nature 537 7620 382 386
152. ZY Huo JF Zhou Y Wu 2018 A Cu3P nanowire enabling high-efficiency, reliable, and energy-efficient low-voltage electroporation-inactivation of pathogens in water[J] J Mater Chem A 6 39 18813 18820
153. R Sa Y Yan Z Wei 2014 Surface modification of aramid fibers by bio-inspired poly (dopamine) and epoxy functionalized silane grafting[J] ACS Appl Mater Interfaces 6 23 21730 21738
154. Y Fu L Liu L Zhang 2014 Highly conductive one-dimensional nanofibers: silvered electrospun silica nanofibers via poly (dopamine) functionalization[J] ACS Appl Mater Interf 6 7 5105 5112
155. NG Rim SJ Kim YM Shin 2012 Mussel-inspired surface modification of poly (L-lactide) electrospun fibers for modulation of osteogenic differentiation of human mesenchymal stem cells[J] Colloids Surf, B 91 189 197
156. B Fei B Qian Z Yang 2008 Coating carbon nanotubes by spontaneous oxidative polymerization of dopamine[J] Carbon 46 13 1795 1797

The manufacturer's authorised representative in the EU is Springer Nature Customer Service Centre GmbH, Europaplatz 3, 69115 Heidelberg, Germany. If you have any concerns regarding our products, please contact ProductSafety@springernature.com

Printed and bound by CPI Group (UK) Ltd, Croydon, CR0 4YY
01/10/2026
02225877-0009